I0074320

GRASSHOPPER
Diversity and Control

The Authors

Dr. Sathe Tukaram Vithalrao [M.Sc., Ph.D., Sangit Vishard, IBT (Seri.), F.I.S.E.C., F.S.E.Sc., F.S.L.Sc., F.I.C.C.B., F.S.S.I.] is presently working as Professor and Head, Department of Zoology, Shivaji University, Kolhapur. He has teaching experience of 29 years in Entomology at University PG department and 15 years in Agrochemicals and Pest Management. He has written 30 books and published 255 research papers in national and international journals of repute. He guided 20 Ph.D. students and completed 6 major research projects (from CSIR, DST, DBT and UGC). He visited Canada (1988), Japan (1988), Thailand (2002, 2004), Spain (2005), France (2005), South Korea (2006) and Nepal (2007) etc. for academic work. He is member of editorial board of eleven prestigious journals. He delivered 35 talks through All India Radio and internal conferences and involved in Doordarshan, S.T.V. and B. T.V. programmes on useful and harmful insects. He published more than 35 popular articles in daily newspapers on insects and sericulture. He got several prestigious awards like "Environmentalists of the Year-2003", "Bharat Jyoti", "Jewel of India", "International Gold Star", "Eminent Citizen of India", "Education Acumen", "Best Educationist", "Eminent Scientist of the Year-2008", "Lifetime Education Achievement", "Lifetime Achievement in Entomology and Insect Taxonomy-2009", Educational Leadership-2011, Asia Pacific International Award-2012 etc. He is also working as Research and Recognition (RR) Committee member for Pune University, Pune; North Maharashtra University, Jalgaon; Shivaji University, Kolhapur and DBA Marathwada University, Aurangabad. He has been awarded several fellowships from different scientific and academic societies. He is Chairman of Maharashtra District Environmental Centre of NESA.

Mr. Bhusnar Appasaheb Ramchandra, M.Sc., Ph.D. is working as Project Scientist in Department of Zoology, Shivaji University, Kolhapur. He has published eight research papers in National and International Scientific Journals/Conferences. He is teaching PG Classes in Pest Management, Department of Zoology, Shivaji University, Kolhapur and life member of Silk Society of India, Nagpur.

GRASSHOPPER
Diversity and Control

— Authors —

T.V. Sathe

Head
Department of Zoology
Shivaji University
Kolhapur – 416 004, M.S.

A.R. Bhusnar

Department of Zoology
Shivaji University
Kolhapur – 416 004, M.S.

2014

Daya Publishing House®

A Division of

Astral International Pvt. Ltd.

New Delhi – 110 002

ISBN 9789351301929

Published by : **Daya Publishing House®**
A Division of
Astral International Pvt. Ltd.
– ISO 9001:2008 Certified Company –
4760-61/23, Ansari Road, Darya Ganj
New Delhi-110 002
Ph. 011-43549197, 23278134
E-mail: info@astralint.com
Website: www.astralint.com

Laser Typesetting : **Classic Computer Services**, Delhi - 110 035

Printed at : **Replika Press Pvt. Ltd.**

PRINTED IN INDIA

Preface

Grasshoppers belong to Order Orthoptera which contains 20,000 described species all over the world. Grasshoppers are medium to large sized, wide spread and diversified taxonomical group of insects. They found associated with various plant species in ecosystem. Grasshoppers are very destructive pests of several agricultural crops and forest plants. Hence, any advance knowledge on their taxonomy, diversity, seasonal abundance and distribution has practical utility in adopting their control measures.

Western Ghats is among 18 biodiversity hot spots. Biodiversity conservation and protection is extremely useful task for sustainable development of a country. In the present text attempts have been made on taxonomical diversity, seasonal abundance, distribution, species richness and check list of the districts Satara, Sangli and Kolhapur of Western Maharashtra. In all, 54 species of Grasshopper have been reported from three districts. Out of which 28 species were highly destructive to crops both in agro and forest ecosystems. 44 species were common and 10 species were rare and 4 species of grasshopper were newly described. Biology and control of grasshoppers is also illustrated .We feel that present work will be useful for students, teachers, scientists, agriculturists and environmentalists.

T.V. Sathe

A.R. Bhusnar

Contents

Chapter 1
General Introduction

India travelled a long way in its drive towards economic growth, modernization of the economy and self reliance of various areas (Dreeze and Sen 1995). India made substantial progress in most indicators of human development over the past fifty years, in an average expectancy.

Water, soil, forestry and biodiversity have direct relation with the environment quality and life style of the animals. Any nation cannot survive without rapid socio-economic development and without resources. Forestry is an important resource from an environment which has direct relationship with the human growth.

Forest is versatile assemblage of biodiversity and ecological system consisting trees and dominated vegetative associations. It is good source of wood, medicine, fuel, fiber, timber grasses and forest products supportive to commercial and industrial activities. It also maintains the ecological balance and life support. Out of 329 million hector of geographical areas of India (Figure 1), forest cover exist only 76.52 million hector of which significant proportion is contributed by lands outside. However, deforested and degraded forest area is quite much greater. Forest cover in 105 districts were more than 33 per cent, 32 districts were within the range of 19-33 per cent and 22 district less than 19 per cent while, 30 districts were without forest covers (FST, 1996).

Due to unplanned and excessive deforestation, a gap between demand and availability of fuel, wood, industrial wood and other forest products is increased. According to the National forest policy the essential forest area is 33 per cent. However, in India, forest area with 40 per cent or above crown cover is only 8.5 per cent.

Deforestation has risen due to excessive felling of trees of timber, over grazing, fire, pests, diseases and clearance of land for cultivation and pastures. Grasshoppers are very bad pests of forest and plain agricultural crops. They skeletonize entire plant by feeding on leaves.

Forests are vanishing at a rate of 17 million hector per year. Deforestation is major cause of loss of biodiversity and species extinction. According to Radhakrishnan (2003) the life forms have totally vanished while some others are categorized as endangered. About 3120 species of plants and 172 categories of animals of our country are listed in the IUCN's red list threatened category.

Tropical, Sub-tropical, Temperate and Alpine types of forest are seen in India (Champion and Seth, 1968). The above groups may be further divided into sixteen sub groups (types), such as tropical dry deciduous, tropical moist deciduous, tropical thorn, tropical wet evergreen, sub tropical pine, sub alpine, Himalayan moist temperate, tropical semi evergreen, montane wet temperate, Littoral and swamp, sub–tropical broad leaved hill, sub–tropical dry evergreen with total percentage of forest area 38.2, 30.3, 6.7, 5.8, 5.0, 4.3,3.4, 2.5, 2.0, 0.9, 0.4, 0.2, 0.2 and 0.1 million hector respectively (Champion and Seth, 1968). According to FAO the annual deforestation rate for India was 0.6 per cent (0.34 million hectors) during 1988 to 1990. Himalayan region, Western Ghats and Northern Eastern forest are the main forests scattered in India.

The Western Ghats (Figure 2) extended from the Southern tip of peninsula from (8pN) to the mouth of river Taphi (21pN) at altitudes between 900 and 1500 m above sea level. However 'Phalghat gap' which is almost the central part of Western Ghats located in Kerala state having very low altitude, less than or about 144/m above the sea level (Joseph, 2004).

The Western Ghats is spread in five states (Figure 2) of India namely Maharashtra, Goa, Karnataka, Kerala and Tamilnadu. At some points is

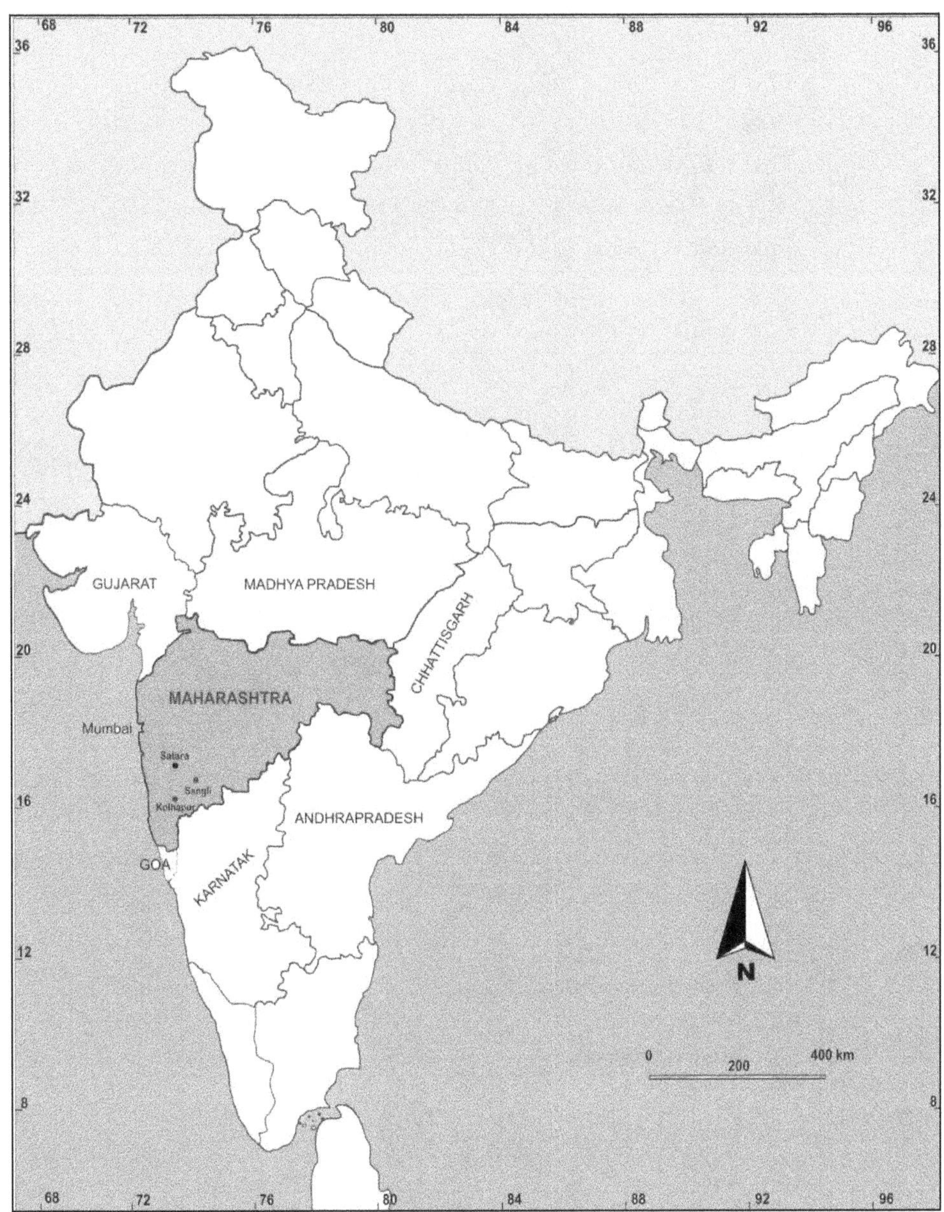

Figure 1: Map of India Showing Maharashtra and Study Area

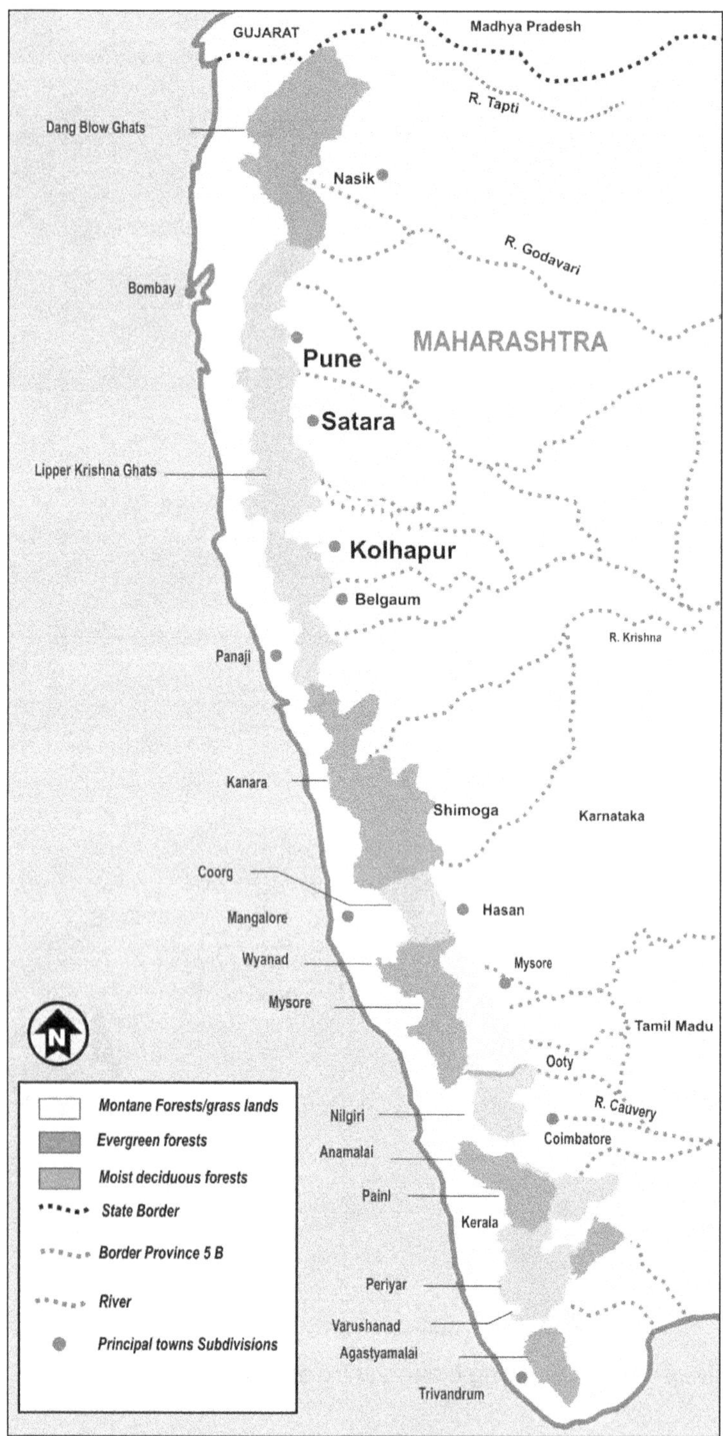

Figure 2: Biographic Subdivisions of the Western Ghats

narrow and very close to the sea. The Western Ghats is one of the biodiversity rich area due to the humid tropical climate, unusable geological stability and evolutionary continuity, which covers 1, 59, 000 sq km area. It is featured by hill terrain and assemblage of discontinuous hills and plateau like Bhimashankar, Mahabaleshwar, Nilgiri and Annamalai.

In Maharashtra Western Ghats is scattered in nine districts namely, Dhule, Thane, Raigad, Pune, Satara, Sangli, Kolhapur, Sindhudurg, and Ratnagiri of which Thane, Raigad, Ratnagiri and Sindhudurg situated at Western slope while, Dhule, Pune, Satara, Sangli, and Kolhapur situated at eastern slope of the Western Ghats.

Biodiversity is the process of correctly identified specimens, checking their distribution to understood their ecological status and process of evolution, vouchering the identified organism and integrating this information into Internationally Accessible Databases.

Probably discovered species of insects from world numbered as 9,50,000, plants 2,70,000, arachnids 75,000, mollusca 80,000, vertebrates 56,000, protozoa 30,000, crustaceans 75,000, other invertebrates 1,20,000 and undiscovered insect species are 89,50,000, plant 3,80,000, arachnids 7,40,000, mollusca 2,50,000, vertebrates 61,000, protozoa 2,10,000, crustacean 1,80,000 and other invertebrates 4,00,000.

17,000 flowering plants species reported from India, out of which more than 4000 occur in the Western Ghats. The biodiversity of lower plants in Western Ghats is equally rich. Roder and Panwar (1988) reported 518 rare or endangered species. The Western Ghats also represents 250 species of orchids, of which 100 are endemic. Yadav and Sardesai (2002) contributed on the flora of Kolhapur district, they reported 2227 species of angiosperms belonging to 1023 genera and 182 families, of which 1620 species of dicots belonging to 788 genera, 149 families, 5 subfamilies and 18 varieties and 607 species of monocot belonging to 235 genera, 33 families and 12 varieties out of which 136 species threatened, 340 genera endemic, 9 genera critically endangered, 22 genera endangered. The Western Ghats is characterized by a very rich variety of fauna. Many species of them are highly endemic. Several reptiles and birds species are confined to only Ghats (Pandharbale and Sathe, 2008). Roders and Panwar (1988) listed many important species and groups of species found in Ghats. The review of literature indicates that Balakrishna and Alexander (1984) studied wild mammals of Western Ghats of Kerala state.

Large numbers of insect species have been recorded from Indian forests. However, little information is available on the characteristics species profile. About 67,000 species of insects have been described from India by 1980 (ZSI, 1983) including forest insects.

In Indian forest insect study was initiated in 1900, prior to the establishment of Forest Research Institute, Dehradun, India. Very large numbers of wood boring beetles of forests have been reported by Stabbing (1940). Beason (1941) worked on several forest insects. He studied about 3378 insects belonging to the Orders Collembola, Ephemerida, Anaplura, Dermaptera, Isoptera, Coleoptera, Hemiptera, Thysanoptera, Thysanura, Hymenoptera, Lepidoptera, Neuroptera, Odonata, Orthoptera, and Diptera. According to Roonwal *et al.* (1950) the forest insects includes over 16000 species they are nearly one quarter of the 6700 species of insects recorded from all habitats in India. Chatterjee and Misra (1974), reported 680 species of parasitoids as natural enemies of insects. Larsen (1987) recorded 299 species of butterflies from the Nilgiri mountain of Western Ghats. A total of 314 species of butterflies have been recorded earlier from the Western Ghats region (Nair *et al.,* 1973). ZSI (1979-1980) studied the insect of salient valley. They reported 242 species of insects from valley. The insect life and ecodevelopments of Western Ghats have been studied by Joseph (1984). He highlighted the role of insects in the Western Ghats ecosystem.

Sathe *et al.* (1986; 1987) studied the fauna of butterflies from Western Ghats of Maharashtra. They reported 33 species of butterflies from Western Ghats. Out of which 8 species were rare. Shivramkrishnan *et al.* (1990) studied Ephimerida from Western Ghats. Sathe and Mulla (1995) reported 17 insect pests feeding on mulberry plants in Amboli from Western Ghats. Gaonkar (1996) listed 330 species of butterflies from the Western Ghats. Satish (1996) recorded 13 species of moths and 16 species of butterflies from Shimoga. Sathe and Pandharbale (1999) described 13 species of Sphingid moths belonging to six genera *viz. Acherontia, Herse, Theretra, Macroglossum, Hippotion* and *Campsogene* from Western Ghats of Maharashtra. Data on Coleopteran is very much limited (Shivramkrishnan *et al.*, 2000). Shivramkrishan *et al.* (2000) also recorded the higher taxa levels in Order Diptera and Odonata. Pandharbale and Sathe (2001) described a new species of Syntomid moth, *Syntomis vitex* from the environment of Western Ghats of Satara district.

Bhoje and Sathe (2003) reported 80 species of butterflies from Radhanagari Wild life Sanctuary. Nagendran and Snija (2004) studied the diversity of higher taxa of aquatic insects in three hill streams of Western Ghats of Southern parts. Recently, Sathe (2004) noted that lady bird beetles like *Menochilus sexmaculata, Coccinella yamini* are others migrating to hilly region of Western Ghats of Maharashtra from Plain region. Pandharbale (2004) described 31 species of moths from Western Ghats of Satara. Out of which 16 species have been newly described.

Sathe and Pandharbale (2005) reported 130 species of moths from the Western Ghats of Satara district. Sathe and Shinde (2008) studied the biodiversity of dragonflies from Koyana dam and surrounded area. They reported 151 species of dragonflies and damsel flies. Out of which 17 species of dragonflies have been newly described. Sathe and Shinde (2006) also studied diversity of butterflies from Western Ghats of Kolhapur district. They reported 80 species of butterflies. Biodiversity of aphids from Western Ghats and Maharashtra have been studied by Sathe (1992). Jadhav and Sathe (2006 a,b). Sathe (1992) reported 28 species of aphids on economic plants from Western Ghats. Similarly, Jadhav and Sathe (2006 a, b) recorded 21 species of aphids from Western Ghats of Poona district. Kawane and Sathe (2011) described new 9 subspecies of Wild silk moths *Antheraea mylita* from Western Maharashtra.

Review of literature indicates that many insects group are attempted from Western Ghats. However, grasshoppers were given very less attention.

Indian agriculture is a backbone of Indian economy. According to Singh and Sadhu (1986), India cultivates a wide variety of food and non food crops. However, the expected yield of the crop is not achieved so far due to the insect pests damage. New hybrid varieties invite pest problems in agricultural crops. Amongst the insect pests, grasshoppers are dominant creatures.

Grasshoppers (Order–Orthoptera) are as an important part of ecosystem and very rich in grassland, forests, deserts, agricultural fields, lawns, wetlands, etc. They are reported from about all habitats, excluding polar region. Grasshoppers play an important role in food chain regulation and nutrient recycling and acts as dominant herbivorous in grassland ecosystem. Majority of the species are tropical but are also well represented in temperate areas. Vertically, they are found from sea level to the high altitudes in the Himalaya (Bhowmik and Rui, 1982; Mehta et al., 2002; Shishodia *et al.,* 2002).

Grasshoppers are one of the largest and diverse groups of insects and they are important food for several birds, reptiles, amphibians and mammals. Most of the grasshoppers are oligophagus and exhibit definite host preferences (Mulkern 1967). On the basis of food preference, grasshoppers are classified as grass feeders (Graminivorous), Forb-feeders (forbivorous) or a mix of the two cambivorous or mixed feeders (Isely, 1944).

According to Paulraj (2009) host plant shifting may occur in grasshoppers when their main host is absent. In recent years farmers have altered cropping patterns and agronomical practices due to urbanization, labour problems and for greater profits. The changing scenario in agriculture is affecting primary consumers like grasshoppers and thereby creating impacts on entire food webs. However, grasshoppers are very bad pests of several agricultural crops. They skeletonize the entire crop. Therefore, it is necessary to study the biodiversity and distribution of grasshoppers in relation to their habitat and host plants.

Biodiversity is the important natural resource of uncountable value. It gives the information regarding the current status of animal evolution, species richness, species abundance and other ecological information from selected areas.

Biodiversity is the richness of an ecosystem (Echrlich and Echrlich, 1981), variety of life and its processes (Keystone 1991), 'Life's endless forms (Darwin 1853) and essence of life (Frankel, 1970). According to Jenkins (1992) measures of biodiversity for particular areas, habitats or ecosystems are largely reduced to a straight forward measure of species richness.

Biodiversity is subdivided into main four types, *viz.*, ecosystem diversity, species diversity, genetic diversity and landscape diversity. A diversity within species is a genetic diversity. Diversity between species is species diversity and diversity of ecological level is ecosystem diversity. (Noss 1992, 1996; Szaro and Sapiro, 1990; Szaro and Swalwasser, 1991; Wilson 1988). Landscape is a heterogenous land area composed of a cluster of interacting ecosystems that is repeated in similar form throughout (Forman and Godron, 1986); Landscape diversity includes the forest, grassland, aquatic and deserts habitats. Biodiversity provides both opportunity and challenges how ecological communities are affected by human activity and environmental perturbations (Kudla *et al.,* 1981).

There has been interest in evaluating the insect richness and diversity of the Indian fauna (Gadagyvar *et al.,* 1990; Murlirangan *et al.,* 2003; Sanjayan 1993). Plant diversity is positively affected by grasshopper diversity. However, grasshopper diversity can be affected by many factors such as host plant availability, type of soil, rainfall, humidity, temperature, latitude-altitude of habitat, predators and other animal interference in habitat.

Grasshoppers belongs to Order Orthoptera. The Order Orthoptera is subdivided in two suborders Ensifera and Califera. Suborder Ensifera includes eight superfamilies and seventeen families while in suborder Califera there are 11 superfamilies and 36 families, out of which four families *viz.* Acrididae, Tetrigidae, Pyrgomorphidae and Tettiigonidae are economically important since the members of these families cause heavy damage to agricultural and forest crops, hence selected for present work.

Under the family Acrididae all locusts are included. Many species of locusts are recorded as serious agriculture pests. They are highly economically important as some of the species cause considerable damage to the agricultural and forest plants (Coleman and Kannan, 1911).

Acridids cause extensive damage to agro ecosystems (COPR–1982). A high population of grasshopper cause rapid loss of vegetation may result in increase runoff, soil erosion and destruction in food sources of many animals. Grasshoppers may be univoltine, bivoltine, or trivoltine. They lay eggs in pods on leaves, grass stems, plant stems, etc. Grasshopper breeding period is highly species specific, mostly depends on the climatic conditions. They breed in very high populations, mostly in monsoon season which is more suitable for grasshopper development.

The weather and soil characteristics are also important to grasshopper populations as they affect oviposition behaviour, embryonic development and egg survival. Soil type, pH, soil texture have also been shown to be important in certain grasshoppers. The effects are probably species–specific (Mukherji and Randell, 1975; Hewilt 1985; Johnson 1989).

Acrididae pass through 4 to 7 nymphal stages for adult. Mated female lay eggs in the upper layer of soil, 7 to 10 cm deep in pods. Incubation period depends on climatic conditions. Newly emerged nymph feeds on young vegetation. The Acridids are mainly phytophagus. After the mating males die soon and after oviposition females too. The warmer months are considered to be most favourable for population of grasshoppers (Bhowmik, 1986). According to Phipps (1970) number of generations in grasshopper, depends chiefly on the size (Weight of gravid females).

Family Pyrgomorphidae is closely related to the family Acrididae. The members of family Pyrgomorphidae are medium sized, more or less granulated, conical headed with antennae inserted between and close to the eyes. They are ensiform type with tegmina pointed at the tip and body dorsoventrally flattened. A mated female can lay eggs in the soil. Newly emerged nymphs feed on the newly emerged vegetation and they molt 7 to 8 times to become an adult. The members of family Pyrgomorphidae are totally herbivorous.

Family Tettigoniidae includes long horned grasshoppers or katytids. They are characterized by having antennae long with numerous segments, long saw or sickle shaped ovipositor and hind femora long narrow towards the tip. Maximum members of Tettigoniidae are active at night. Stridulation is an important character of Tettigoniidae. Stridulatory mechanism mostly used for attracting the female for mating and warning to other animal when they are in danger. Stridulatory mechanism in Tettigoniidae is mostly carried out at night. Auditory organ present near the base of fore leg tibia.

The family Tettigoniidae is a heterogeneous group of grasshoppers with more than 1120 recognized genera and 6800 species and is the largest family within the Orthoptera.

Members of Tettigoniidae lay eggs on the plant stems or ventral side of plant leaves in groups, mostly in post monsoon period. Long horned grasshoppers hatched in monsoon season. Newly emerged nymphs of grasshopper feeds on the developing barks, grain ovaries, flowers, small insects etc. They moult 7-8 times. Predatory species are also from this family *Hexacentrus unicolor, Mecopoda elongata, Euconocephalus incertus, Phenoroptera gracilis, Conocephalus maculatus* observed feeding on lepidopterous larvae, aphids, mealy bugs, thrips and other small insects.

The family Tetrigidae contains the grouse locusts. They are small, mostly blackish brownish, whitish and greenish coloured. The Tetrigidae recognized by their special type of pronotum which is elongated and hard, tegmina very short, wings are well developed, antennae short hair like, fore and mid legs delicate, claws without arolium, hind legs adaptive, tarsus three segmented and eyes rounded, mustard like. Maximum Tetrigids live in dwelling of soil and below the shedding of vegetation and other organic matter of soil surface. Mostly they found near the fresh of water bodies. They emerge in month of June-July and mature in September-October; after maturation their mating take place, mated females lay eggs in soil during months of September-October. During months of June-July, they emerged in huge number in the surroundings of the fresh water bodies. Both nymphs and adults of grouse locust feed on mosses, algae and other vegetations. They are also important agricultural pests.

Locust attack is world wide as natural hazards made by some species of family Acrididae. They migrate from one region to another region in a very huge numbers. While crossing the regions they cause severe damage to agricultural crops. Near about 70-80 per cent damage is recorded to agricultural crops by the Locust. In India Locust attack is reported from Northern region, mostly in Gujarat, Rajasthan and Punjab region. However, Rajasthan is very good breeding place for migratory Locust.

Migratory Locust lives in three stages *i.e.* solitary phase, gregarious phase and congregens phase. In solitary phase, they live in small area with a small group. In gregarious phase they live in very large group and migrate from one region to another region. Congregens phase is the phase between solitary and gregarious phase. Migration found only in gregarious phase. In solitary phase Locust egg hatching is normal but in gregarious phase eggs go in long diapause, when suitable climatic conditions are available they hatch in very high number and cause a locust attack. The reasons of migration are not perfectly known to the science but, mostly they migrate due to unsuitable climatic conditions, lack of food and in search of suitable breeding areas.

Orthacanthacris aegypti, Schistocerca gregaria, Calliptamus italicus, Calliptamus barbarus, Calliptanns ictericus, Dosciostaurus moroccanus, Locusta migratoria, Locusta migratoria migratoria, Locusta migratoria tossia, Locusta migratoria capito, Locusta migratoria migratoroides, Locusta migratoria manilensis, Odedaleus nigrofasciatus, Locusta pardalina, Schistocera paranensis, Nomadacris septemfaciata, Cyrtacanthacris succincta, Melanoplus spretus, Cyrtacanthacris tatarica, Patanga suncincta, Anacridium flavescence, etc. are the migratory locusts reported from all over the world. However little is known on Indian species of locusts.

Kolhapur, Satara and Sangli districts of Maharashtra are very good agricultural and forest spots in India. Secondly, Western Ghats is one of the hot spot of the world in biodiversity and grasshoppers are quite destructive insects of various crop plants. Hence, the present study will add great relevance in conserving and protecting the biodiversity of both, plants and animals, not only from Western Ghats but also from plain region of Maharashtra.

Chapter 2

Review of Literature

At International scenario, several workers have been worked on grasshopper. Serville (1839) studied the Orthoptera from Paris. Walker (1871) prepared a catalogue of the specimen of Dermaptera and Saltatoria from the collection of British Museum London. while, Joern (1975) studied the Grasshopper affinities and habitat relations in the Solitario. Bailey (1979) made a review of Australian Copiphorini (Orthoptera : Tettigoniidae : Conocephalinae). Joern (1982) studied distribution, densities and relative abundance of grasshoppers (Orthoptera : Acrididae) from Nebrasko Sandhills Prairiie. Ingrisch (1989) also studied Acrididae from Thailand and adjacent regions.

Kemp *et al.* (1990) reported the pattern of vegetation and grasshopper community composition. Jago (1998) demonstrated the world wide magnitude of Orthoptera as pests. Ronwell (1998) studied grasshoppers of Costa Rica with respect to survey influencing their conservation and survival. Sergeev (1998) studied the Orthopteran biological diversity related to Landscape change in temperate from Elurasia. Stewart (1998) detected the non-target grasshoppers as indicators of the side effects of chemical locust control from Karoo., South Africa. Suhali *et al.* (1999) studied taxonomy of subfamily Hemiacridinae while, Cigliano *et al.* (2000) reported grasshopper (Orthoptera : Acridoidea) species diversity from the Pumpas Argentina. De. Wysiecki *et al.* (2000) Studied the Grassland and Shrub land grasshopper community composition in northern La Pampa Province, Argentina. Lockwood and sergeev (2000) reported the comparative biogeography of grasshoppers (Orthoptera : Acrididae) in North America and Sineria.

Muhammad Azhar, *et al.* (2000) worked on Biosystematics of Genus *Formosatettix tinkham* (Tetrigidae : Orthoptera) of paddy tract from Punjab, Pakistan while, Andersen *et al.* (2001) worked on grasshopper biodiversity and bioindicators in Kakadu National Park, Australia. Ingrich and Garaj (2001) reported the Orthopteroid insects from Ganesh Himal, Nepal. Gebeyehu and Samways (2001) studied the conservation refugium value of a large mesa for grasshoppers in South Africa while, Floren *et al.* (2001) studied the diversity of Orthoptera from Bornean lowland rain forest trees.

Preurling (2001) worked on environmental conservation and locust control, possible conflicts and solutions. Samways and Keuzinger (2001) studied the vegetation, angulate and grasshopper interactions inside Vs. outside from African Savanna game park. Suhali *et al.* (2001) reported some Acridid grasshoppers belonging to subfamily Cyrtacanthacridinae (Acrididae: Orthoptera) from Pakistan.

Cigliano *et al.* (2002) studied Grasshopper (Orthoptera: Acrididae) community composition and temporal variation in the Pamps, Argentina. Gebeyehu and Samways (2002) reported Grasshopper assemblage response to a restored national park (Mountain Zebdra National Park South Africa) while, Ingrisch (2002) reported Othroptera from Bhutan, Nepal and North India. Kruess and Tscharntke (2002) studied Grazing intensity and the diversity of Grasshopper. Lanjar Abul Ghain *et al.* (2002) reported the occurrence and abundance of Grasshopper species on rice from, Pakistan.

O' Neill *et al.* (2002) studied sweep sampling technique, relative abundance and community composition of grasshoppers (Orthoptera : Acrididae). Bieringer and Zulka (2003) reported shading out species richness, edge effect of a pine plantation on the Orthoptera (Tettigoniidae and Acrididae) assemblage of an adjacent dry grassland.

Mass (2003) studied the new and less known Orthoptera (Insecta) from the Island of Socatra (Yemen) while, Showler (2003) studied the importance of armed conflict to desert locust control, 1986-2002. Forsberg (2004) worked on micro habitat, phenology and diversity of Orthoptera from seminatural pasture, from Uppsala while, Gardiner and Hill (2004) worked on the feeding preferences of *Chorthippus parallelus* (Orthoptera: Acrididae). Heller *et al.* (2004) studied the *isophya* species from Central and Western Europe (Orthoptera : Tettiigonoidae : Phaneropteridae).

Smith *et al.* (2004) prepared key to the grasshoppers (Orthoptera : Acrididae) of Florida. Kim and Kim (2004) made a taxonomic study of Korean Tetrigidae (Orthoptera : Caelifera : Tetrigidae). Agabiti *et al.* (2005) studied the grasshoppers (Insecta-Orthoptera) of a sub-mediterrean zone of the Trentino region (North-East Italy). Gardiner *et al.* (2005) reviewed the methods frequently used to estimate the abundance of Orthoptera in grassland ecosystem.

Joern (2005) studied the disturbance by fire frequency and Bison grazing modulate grasshopper assemblage in tall grass Praire. Lecoq (2005) studied desert locust management from Ecology to Anthropology. Schmitz (2005) studied grasshoppers from forest ecosystems. Smith and Capinera (2005) studied mandibular morphology of some Floridian grasshoppers (Orthoptera : Acrididae). Yin and Yin (2005) described of two new species of *Stenocatantops* (Orthoptera : Acrididae : Catantopinae) from Taiwan with a key to known species of genus.

Ahnesjo and Forsman (2006) worked on habitat selection by Pygmy grasshopper colormorphs and interactive effect of temperature and predator avoidance. Song (2006) described new species *Schistocera cohmih* and redescribed *S. socrro* (Dirsh) (Orthoptera: Acrididae: Cyrtacanthacridinae) from Mexico. Badenhausser *et al.* (2007) estimated the Acrididae densities in grassland habitats and compared between presence-absence and abundance sampling designs. Batry (2007) studied effect of local and landscape scale and cattle grazing intensity of Orthopteran assemblages of hungarian Great Plain. Brust *et al.* (2007) reported immersion tolerance in rangeland grasshoppers (Orthoptera: Acrididae). Hill (2007) worked on grasshopper (Orthoptera: Romaleidae: Acrididae) fauna of black Belt prairie remnats in Alabam and Mississippi. Jonas and Joern (2007) studied grasshoppers (Orthoptera : Acrididae) communities respond to fire, bison grazing and weather in North American tall grass Prairie. Nagy *et al.* (2007) tested effectiveness and selectivity of three sampling methods frequently used in Orthopterological field studies. Spungis (2007) studied the fauna and ecology of grasshoppers

(Orthoptera) in the coastal dune habitats in Ziemupe nature reserve, Latvia. Steck *et al.* (2007) worked on conservation of grasshopper diversity in a changing environment. Zahan *et al.* (2007) studied low density, cattle grazing and arthropod diversity of Abandoned Wetland while, Amedegnato and Devriese (2008) reported the Global diversity of true and pygmy grasshoppers (Acridomorpha : Orthoptera) in freshwater. Lockwood and Lockwood (2008) studied the population ecology of grasshoppers from North America.

Pisica *et al.* (2008) collected preliminary data regarding the Orthoptera (Insecta: Orthoptera) and specific diversity from Bucharest metropolitan area. Trewick and Morris (2008) gave the taxonomic status of some New Zealand grasshoppers. Branson and Sword (2009) reported grasshoppers affecting on native plant diversity and abundance in grassland dominated by the exotic grasses. *Agropyron cristatum*. Lin (2009) discovered a new genus *Paraxantia* gen. nov., with description of four new species (Orthoptera : Tettigoniidae : Phaneropterinae) from China while, Marini *et al.* (2009) studied response of Orthopteran diversity to abandonment of semi natural meadows. Massa (2009) prepared a checklist of Orthoptera from Libya. Rentz *et al.* (2009) studied Australian Tettigoniidae: (Orthoptera : Tettigoniidae : Phyllophoridae). Similarly, Massa (2009) studied new and less known Orthoptera (Insecta) from the island of Socotra(Yemen).

Lecoq *et al.* (2010) studied the phase-dependent colour polyphenism in field populations of red locust nymph (*Nomadocris septemfasciata* Serv.) from Madagascar. Mukhtar *et al.* (2010) reported biodiversity and occurrence of grasshoppers (Acrididae : Orthoptera) of Quetta Division, Balochistan. Olfert *et al.* (2010) worked on the application of General circulation models to assess the potentials impact of climate change on potential distribution and relative abundance of *Melanoplus sanguinipes* (Fabricius) (Orthoptera:Acrididae) in North America. Song (2010) studied density-dependent phase polyphenism in Nonmodel locust.

Alexander (2011) studied the behavioural ecology and biography of Locusts and grasshoppers. Branson (2011) studied the relationship between plant diversity and grasshopper diversity and abundance from the little Missouri National Grassland. Latchininsky (2011) reported the Locust and grasshoppers: behaviour, ecology and biogeography while, Lockwood

(2011) reported the ontology of grasshoppers. Nutio *et al.* (2011) studied grasshopper response to reduction in habitats area as mediated by subfamily classification and life history traits while, Sergeel (2011) studied the distribution patterns of Grasshopper from Boreal Zone, Russia.

Dominant workers on Orthopteran fauna of India refer to Kirby (1914), Chopard (1969), Bhowmik (1986), Shishodia (1991 to 2010) and Chandra (2003, 2007) etc.

Cotes (1891) studied the Locust of North-Western India *Acridium peregrinum*. Boliver (1899) worked on the Les Orthopteres in Joseph college, Trichinopoly. Kirby (1914) (Orthoptera : Acrididae) reported the fauna of British India, including Cylon and Burma. Uvarov (1925) recorded nine species from Central Province. Hubbell (1932) made a revision of the pure group of the North American Genus *Melanoplus*, with remarks on the Taxonomic value of the Cyrtacanthacrinae (Orthoptera–Acrididae). Dutt (1947) studied on the Acytological investigation of the Genus *Phloeba* (Acrididae). Uvarov (1955) studied the ecology of Locusts and Grasshoppers of the old World. Tandon *et al.* (1976) reported Orthoptera from Kanha National Park, while, Dwivedi (1978) studied the energy flow of grasshopper *Conocephalus pallidus* in Bhata grassland, Madhya Pradesh. The occurrences of Metacentrics in south Indian grasshoppers were studied by Aswanthanaryana *et al.* (1981). Roonwal (1981) reported 10 species of Orthoptera from Central India, while Julka *et al.* (1982) studied ecology of the grasshoppers (Orthoptera: Acridoidea) from din Solan (A.P.). Paranjpe and Bhalerao (1984) worked on bioecology of pigmy locust, *Potua sabulosa* Hancock. (Tetrigidae:Orthoptera). Bhowmik and Halder (1984) studied the distribution of little known species of Acrididae (Orthoptera:Insecta) from Western Himalayas (Himachal Pradesh). Paranjape and Bhalerao (1984) biosystematic studied and ethoecological studies on family Tetrigidae (Orthoptera).

Tandon *et al.* (1988) have taken some observations on the field biology and ecology of some grasshoppers (Orthoptera: Acridoidea) near Calcutta while, Bhowmik (1990) studied the Indian species of the genus *Aulacobothrus* Bolivar (Orthoptera:Acrididae). Bhowmik *et al.* (1990) Contributed to the Acridid fauna (Orthoptera) of North-Eastern states of India. Dwivedi (1990) reported bioenergetics of the grasshopper *Catantops Pinguis innotabilis* (Walker) from Bilaspur (Chhatisgarh). Jago and Bhowmik

(1990) discovered a new species of Oediipodine Grasshopper, *Sphingonotus orissaensis* from Eastern India (Orthoptera : Acrididae). Shishodia and Mandal (1990) made some new records of Orthoptera (Insecta) from the Nagarjun Sagar Tiger Reserve Forest, Andhrapradesh, India. Shishodia (1991) reported eight species of grouse locust from Kanha National Park.

Shishodia (1991) studied the Tetrigidae of North Eastern India. Singh and Singh (1992) recorded *Cyrtacanthacris tatrica* (Linn.) as a pest of soybeans pods from Madhya Pradesh. Shishodia (1995) reported 15 species belonging to six families of Orthoptera from Indravati Tiger Reserve while, Tandon *et al.* (1995) reported 23 species belonging to four families of Orthoptera from Kanha Tiger Reserve.

Ingrisch and Shishodia (1997) discovered a new species of *Agraeciini* from Northeast India (Ensifera: Tettigoniidae). Ingrisch and Shishodia (1998) again discovered one more new species. They made records of Tettigoniidae from India. Shishodia (1999) recorded 22 species of Orthopteran insects from Patalkot, Madhya Pradesh. Ingrisch, Sigfried and Shishodia (1997) discovered a new species of *Araeciini* from Northest India (Ensifera: Tettigoniidae) while, Ingrisch and Shishodia (1998) discovered additional new species from Tettigoniidae. (Ensifera). Ingrisch (1998) prepared a monograph of the oriental Agraciini (Insecta : Ensifera : Tettigoniidae) and done Taxonomic revision; Phylogeny, Biography, ostridulation and development. Sharma *et al.* (1999) studied the population outburst of *Mecopoda elongata* Linn. (Orthoptera : Tettigoniidae) around Pune, Maharashtra. Similarly, Shishodia (2000) studied Orthopteran (Insecta) fauna of Andaman and Nicobar Islands. Shishodia (2000) recorded 77 species of crickets and grasshoppers from Bastar, Chhatisgarh.

Shishodia *et al.* (2002) recorded 39 species of Order–Orthoptera from pong dam wetland, Kangra, Himachal Pradesh, India. Chandra (2003) reported 121 species of Orthoptera from Kanha National Park. Kandibane *et al.* (2003) studied the Orthopteran diversity in irrigated rice ecosystem from Madurai, Tamil Nadu. Ingrisch and Muralirangan (2003) discovered a new species of *Himertual* (Orthoptera : Tettigoniidae) from Tamilnadu (India).

Priya and Narrendrah (2003) prepared a checklist and key of the genus of short horned grasshopper (Orthoptera : Acrididae) from Kerela.

Kandibane *et al.* (2004) studied the Orthopteran diversity in irrigated rice ecosystem from Madurai, Tamil Nadu. Similarly, Khalid *et al.* (2004) studied the Tetrigidae (Orthoptera) from Azad Jammu and Kashir while, Millemse and Ingrich (2004) discovered a new genus and species of Acrididae from South India namely, *Nathanacris quadrimaculata*. Thakur *et al.* (2004) reported the Orthopteran diversity of Roper Wetland Punjab (India). Willemse and Ingrisch (2004) discovered a new genus and species of Acrididae from South India (Orthoptera : Acridoidea).Yack (2004) studied the structure and function of auditory chordotonal organs in insects. Met al. (2005) made a survey of short horned grasshoppers (Acrididae) from Dakshina Kannada district of Karnataka. Mayya *et al.* (2006) prepared list of Orthopteran fauna of the Gibbon Wildlife sanctuary, Assam and reported 25 species of Orthoptera. Shishodia (2006) recorded 21 species of Orthopteran insects from Rewa (Madhya Pradesh). Chandra *et al.* (2007) prepared a checklist of Orthoptera from Madhya Pradesh, Chhatisgarh. Mathew (2007) studied the fauna of Neyyar Wildlife Sanctuary, Kerela (India). Mandal *et al.* (2007) published a pictorial handbook on Indian short horned grasshopper pests (Acridoidea : Orthoptera).

Kanade *et al.* (2008) made vegetation survey of Chandoli National Park from North, Western Ghats of India while, Ananthaservi *et al.* (2009) reported Acridid (Orthoptera) fauna of agricultural ecosystem from some southern district of Tamil Nadu. Chandra and Gupta (2009) collected the Orthopteran fauna from Veerangana Duravati Wildlife Sanctuary, Madhyapradesh, India. Gupta and Chandra (2009) studied Orthopteran of Madhav National Park, Shivpuri, Madhyapradesh, India. Paulraj *et al.* (2009) studied the distribution of grasshoppers (Insect : Orthoptera) with respect to different host plants and habitats in two district of Tamil Nadu, India, they reported 33 species of Orthoptera. Saha and Haldar (2009) declared the Acridids as indicators of disturbance in dry deciduous forest of West Bengal, India. Shishodia and Gupta (2009) prepared a checklist of Orthoptera (Insecta) of Himachal Pradesh, India. Branson (2010) studied the relationship between plant diversity and grasshopper diversity and abundance in the Lihtle Missouri National grassland. Chandra *et al.* (2010) prepared a checklist of Orthoptera (Insecta) from India. Shenthilkumar (2010) made checklist of Orthopteroids in Kaziranga, National Park, Assam, India. Similarly, Shishodia (2010) also prepared checklist of annotated Orthopterra (Insecta) from India. Thakur and Thakur (2011) studied diversity from Sukhna Wildlife Sanctuary Chandigarh, India.

Very recently, Bhusnar and Sathe (2011) studied the diversity of grasshopper from Kolhapur district Maharashtra, India. Bhusnar and Sathe (2012) also studied the ecology of grasshoppers (Order- Orthoptera) Kolhapur district, Maharashtra, India. The review of literature indicates that little information is known about grasshopper from Western Maharashtra. Hence, present work is carried out.

Chapter 3
Materials and Methods

Materials and methods is very important part in the research. Materials help to technical support and methods help in getting good results. A minor change in the method may give the major change in result it may be positive or negative. Therefore, materials and methods have uncountable value in the research.

Following materials and methods were used for studying grasshoppers' diversity.

1) Insect Collecting Net (Figure 5)

Insect collecting net made up of iron handle 70 cm long, circular iron ring of 22 cm diameter covered with nylon net of length 60 cm, forming a bag like structure was used for collecting the grasshopper species.

2) Plastic Containers (Figure 6)

2 lit capacity plastic containers were used for carrying and keeping temporary the collected grasshoppers.

3) Specimen Bottles (Figure 7)

Size 2.5 x 6 cm and 2 x 6 cm (diameter and length) were used for preserving the small grasshoppers.

4) Camel Hair Brushes (Figure 8)

Camel hair brushes No. 4 and 8 have been used for cleaning the preserved specimen.

5) Spreading Board (Figure 9)

Insect spreading boards were used for setting the grasshoppers and for better spread of wings.

6) Oven (Figure 10)

Oven of size 4 x 3 feet (height and width) have been used for drying the grasshopper specimen in 60°C temperature at 1 hr.

7) Insect Storage Box (Figure 11)

Insect storage box size 30 x 45 cm was used for keeping dried grasshoppers.

8) Hand Lens (Figure 12)

Hand lens 10 cm diameter used for study the morphological characters of grasshopper species.

9) Glass Cage (Figure 13)

Glass cage of size 30 x 30cm (height and width) have been used for studying the host and feeding potential of grasshoppers.

10) Compound Microscope (Figure 14)

Simple compound microscope with 10X and 45X objects was used for morphological and taxonomical studies of grasshopper species.

11) Photographs (Figure 15)

Photographs of grasshoppers species and their various body parts have been taken by using photographic camera.

12) Slides and Cover Slips (Figure 16)

Slides and cover slips were used for preparing a slide of body parts of grasshopper such as antenna, wings, mouth parts, etc.

13) Entomological Pins (Figure 17)

Entomological pins of No. 7 have been used for pinning the grasshopper species.

14) Card Sheet Paper

White coloured card sheet papers were used for carding the small grasshopper species mostly Tetrigidae members.

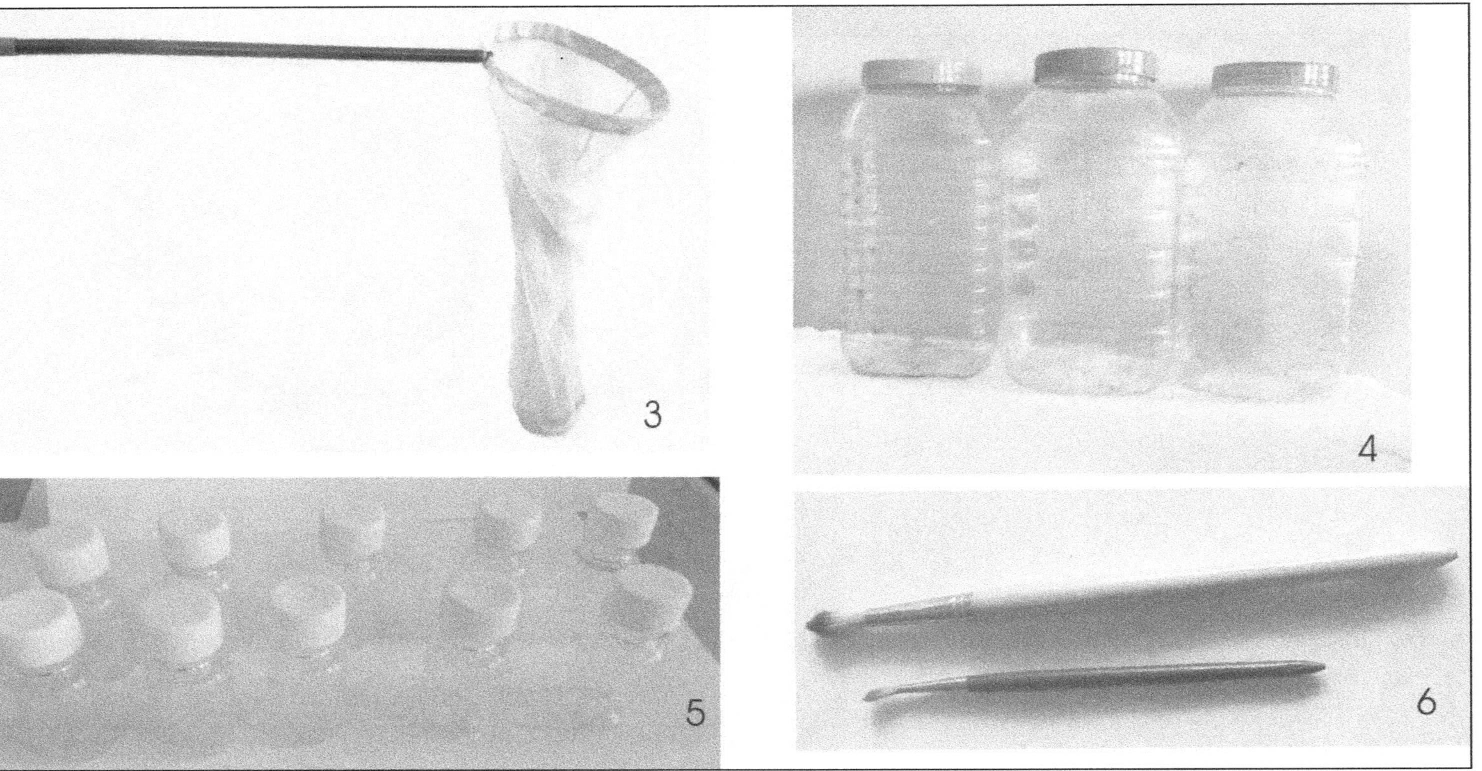

Plate 1: Figure 3: Insect collecting net; Figure 4: Plastic containers; Figure 5: Specimen bottles; Figure 6: Camel hair brushes

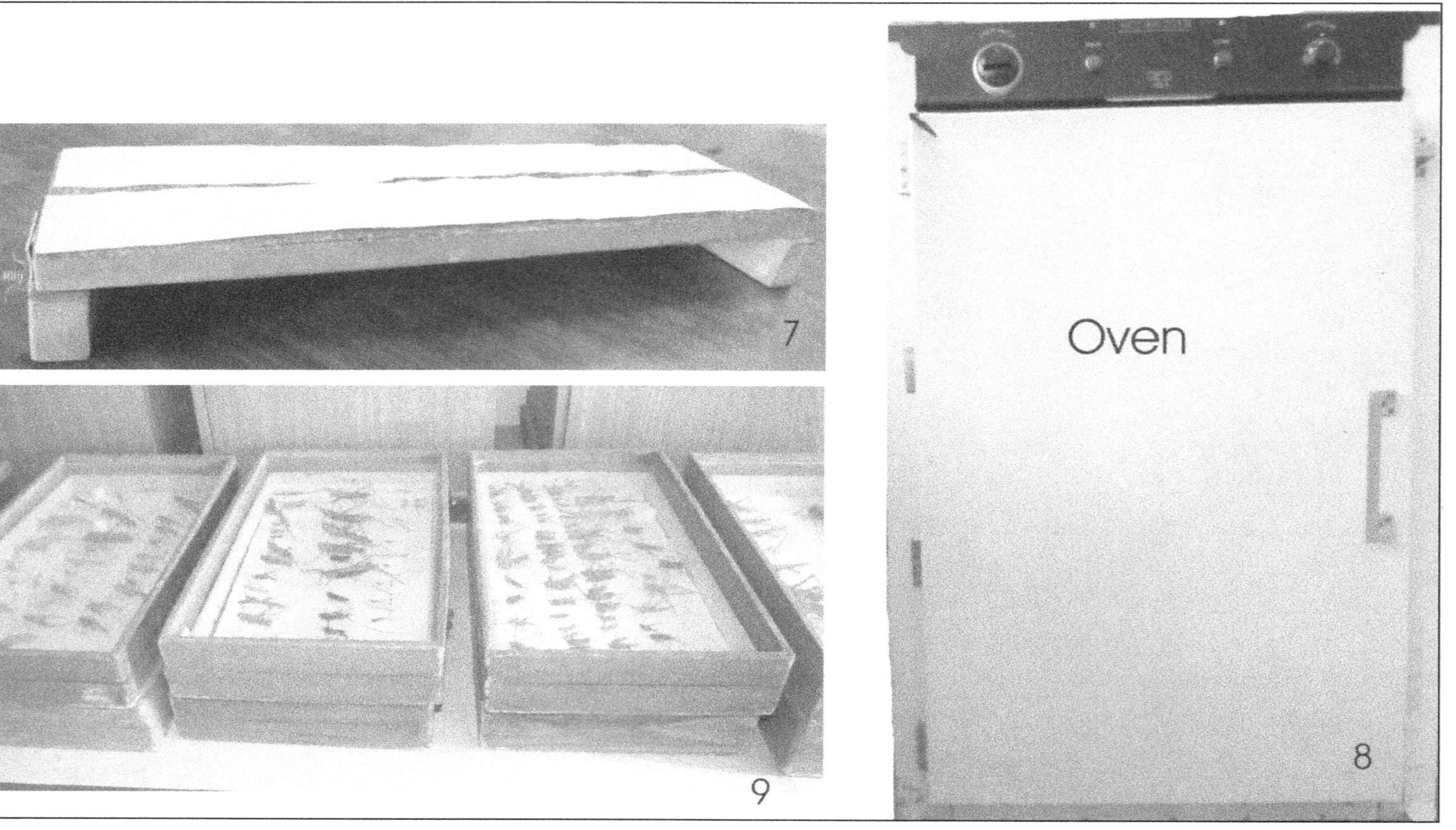

Plate 2: Figure 7: Spreading board; Figure 8: Oven; Figure 9: Insect storage boxes

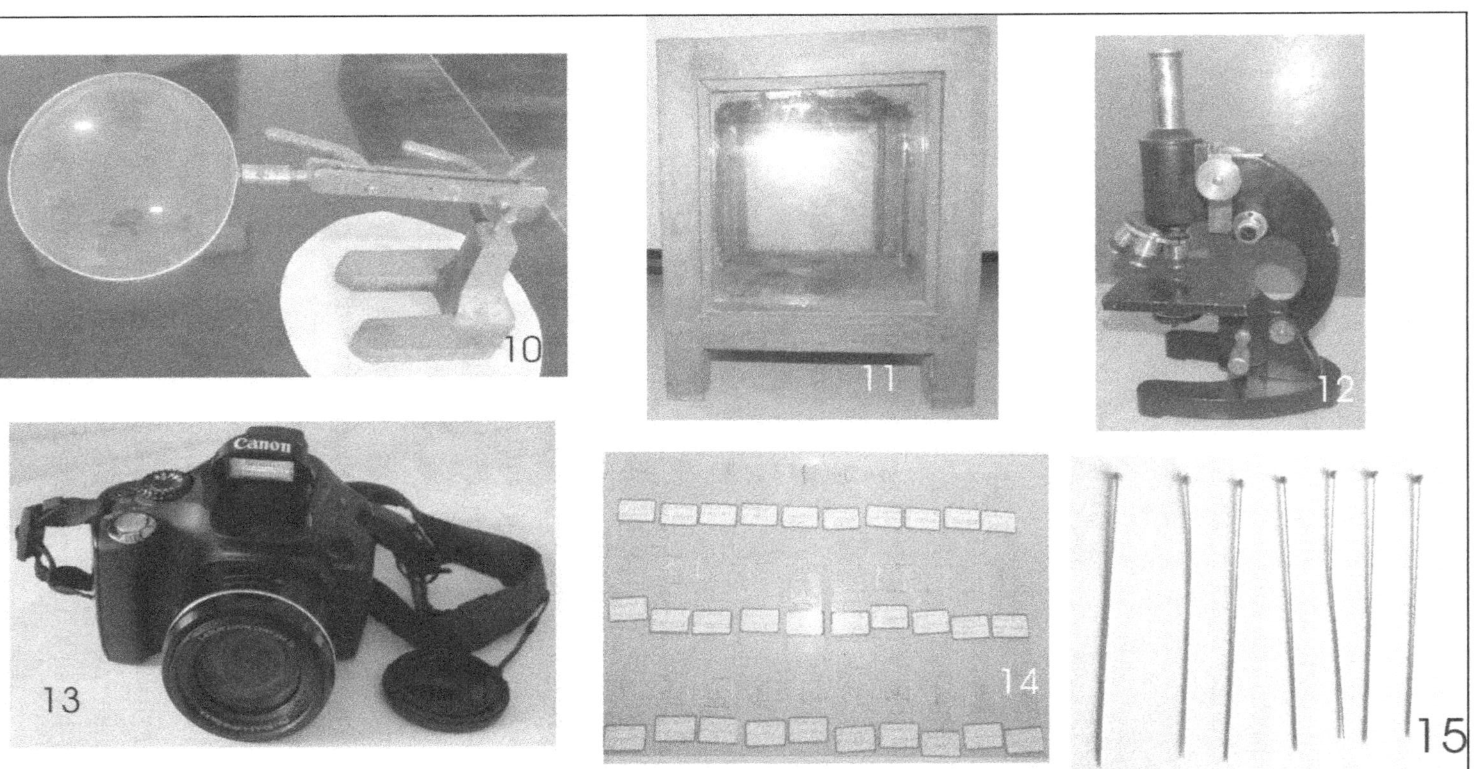

Plate 3: Figure 10: Hand lens; Figure 11: Glass cage; Figure 12: Compound microscope; Figure 13: Camera (Canon SX 30); Figure 14: Slide; Figure 15: Entomological pins

15) Chemicals

Following chemicals were used for preserving the insects.

a. 70 per cent Alcohol for wet preservation.

b. Chloroform for killing the grasshopper specimen.

c. Naphthalene balls were used for safety preservation and avoiding the fungal infection and damage to insects.

Methods

Grasshoppers species were collected from various agro and forest ecosystems of Satara, Sangli and Kolhapur districts. Studies were made morphology and taxonomy of the species by using hand lens and compound microscope. Later photography was made using Canon Power shot (SX30) photographic camera with 35X optical zoom. Thus, collected common species of grasshoppers were killed in killing jar by using chloroform. Killed grasshoppers species were settled on spreading board in a scientific manner and pinned and then dried in oven at 60° c temperature upto 1 hr. After observation, rare and live species of grasshoppers were released where from they were collected. Dried specimens preserved in insect storage box. 4 to 5 naphthalene balls were also used for avoiding the fungal and other infections into storage box.

Wet Preservation Method

Grasshoppers species were collected from various agro and forest ecosystems of study area. Thus, collected common grasshoppers species killed in killing jar by using chloroform. Then they kept in 2 lit capacity plastic containers filled with 70 per cent alcohol and then closed.

Specimen identification and statistical methods in relation to population dynamics are given in respective chapters.

Study Area

The study area shown in Figures 1 to 4 and 18 to 25 is from three districts of Western Maharashtra namely Kolhapur, Sangli and Satara.

Kolhapur district (Figures 16 and 17) has an area of 7585 sq km with total forest coverage of about 1672 sq km. Out of which 563 sq km is reserved and 417 sq km is a protected forests and total forest area is about

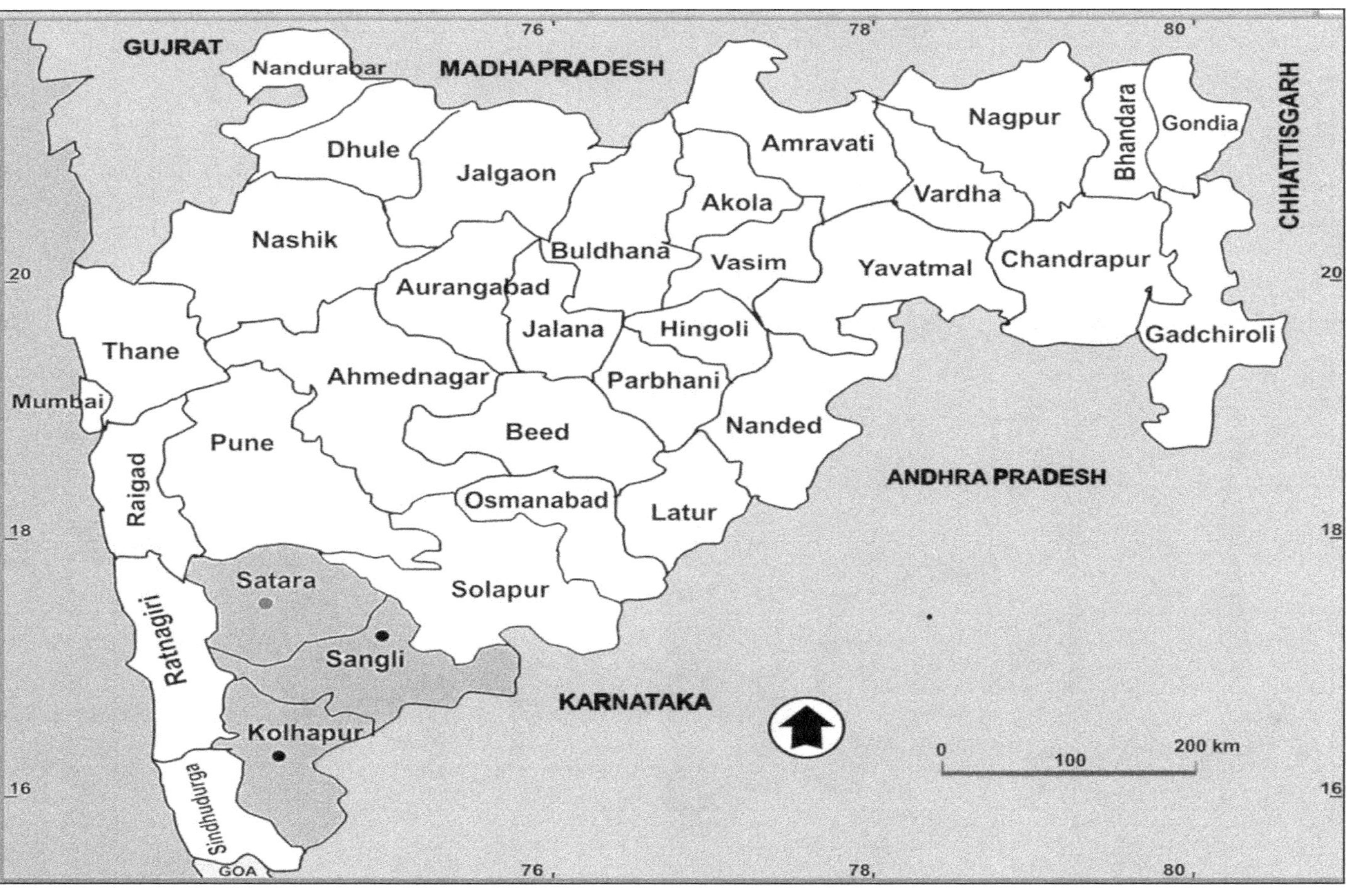

Figure 16: Map of Maharashtra showing study area (Western Maharashtra Districts: Sangli, Satara and Kolhapur)

Figure 17: Map of Study Area (Districts: Kolhapur, Sangli and Satara) showing collection spots)

Plate 4: STUDY SPOT; Figure 18: Western Ghats forest; Figure 19: Study spot Shahuwadi; Figure 20: Amba Ghats tank; Figure 21: Valleys and Hills of Western Ghats (Vishal Gad region)

Plate 5: Figure 22: Study spot Koyna; Figure 23: Study spot Dajipur; Figure 24: Study spot Gadhinglaj; Figure 25: Study spot Bhudhargad

22 per cent. Kolhapur district contain subtropical evergreen and dry deciduous forests.

The subtropical evergreen forest is characterized by the trees such as Anjan, Jack fruit, Mango, Hirada, Jambhul. Semi evergreen and moist deciduous forests are characterized by having Mango, Biba, Ain, Arjun, Kinnai, Umber, Kumbhi, Kinjal, Bhava, Asana, Nana and other while, dry deciduous forests have been classified as reserve and protected forests.

Kolhapur district is mainly divided in three parts namely Eastern ranges, Central ranges and Southern ranges. Eastern and Central ranges have black soil and at some places fertile land. The western ranges are mostly hilly, formed from 'lava' and with red soil and thick forest coverage. Panchganga, Warna, Bhogavati, Dudhganga, Hiranyakeshi, Vedganga and Ghataprabha are main rivers which flows east through Western Ghats.

Temperature in summer rises upto 41.68pC in month April and down up to 14.44pC during the month December. Average humidity of district is 58 per cent. The Kolhapur district receives rainfall from South West monsoon on an average 1600 mm. The rainy season is from June to October, rainfall varies from about 600 mm in Shirol Tahsil in East and Gaganbavada receives 5000 mm average rainfall in West.

Kolhapur district contain 12 tahasils namely Radhanagari, Gaganbavada, Ajara, Shahuwadi, Chandgad, Gadhingalaj, Bhudergad under hilly region of Western Ghats and Karveer, Panhala, Kagal, Shirol comes under eastern part of district which are very important from the view point of agricultural crops.

In present study Amba ghat, Radhanagari, Gaganbawada, Ajara, Chandgad, Gadhingalaj have been selected for Grasshopper diversity. The cash crop sugarcane is highly cultivated, whereas cereals crops like, paddy, maize, wheat, jowar; vegetables like brinjal, tomato, cabbage, cucumber, chilli, etc., pulse crops like cow pea, red gram, green gram, soyabean cultivated in eastern part of district. Fruit crops like mango, coconut, jambhul, jackfruit, cashew nut, guava, etc. and Narcotics and Spices crop like curcuma, ginger and tobacco are also important components for cropping pattern.

Sangli district (Figure 16 and 17) is situated between 16° 45′ and 17 ° 30′. North latitudes and 73° 42′ and 75° 40′ East longitude and bounded on East by Bijapur district of Karnataka State., the West by Ratnagiri and South by Kolhapur (MS) and Belgaum district of Karnataka State. Satara and Solapur lies on the North boundaries of the district. Sangli district falls partly in Krishna basin and partly in Bhima basin and have two drain systems on the basis of topography, climate and rainfall.

Western hilly area of Shirala tahsil with heavy rainfall, the basin of Krishna, Warna, Yerala comprising of Walwa tahsil and Eastern drought porne areas which comprises eastern part of Miraj and Tasgoan tahsil, North estern part of Khanapur and whole of Atpadi, Kavate Mahankal and Jath tahsil. The climate is hotter and drier towards the east and humid towards the west, temperature ranges between 31.1°C in July to 41.5°C in April and 10.3°C in December to 21.5°C from April to June. There are about 76 major and minor irrigation projects in this district.

Study spots in Sangli district refers to Sangli- Miraj, Battis Shirala, Islampur, Tasgoan, Jath, Chandoli, Sagareshwar, Vita.

The district Satara (Figures 16 and 17) has an area of 10,492 sq km, with a part of deccan plateau between latitude 17p 5′ and 18p 1′ N, longitude 73p 33′ and 74p 74′ E. It is bounded by Sangli at southern and Pune at northern side and subdivided into eleven tahsils namely Patan, Satara, Medha, Wai, and Mahabaleshwar which comes under hilly region. Ajinkyatara, Kas, Mahabaleshwar, Panchgani, Yavteshwar, Ghatmatha, Salpani hills, Vankusawade and Bhairavgad hills comes under Satara district and are on the altitude of 900 to 2150 m from sea level. The popular Koyana dam and Koyana Wild Life sanctuary are also part of Western Ghats of Satara. In present study Satara, Ghatmatha (Koyana), Sajjangad, Mahabaleshwar, Karad, Wai, Lonand, Panchganihave been selected for Grasshopper diversity studies. Since they are important in the view point of rich flora and fauna.

In the plain region of Satara district sugarcane, paddy, wheat, maize, sorghum, soyabean, groundnut, red gram, green gram, etc. are important crops cultivated. Vegetables like lady finger, tomato, watermelon, cucumber, etc and fruits plants like coconut, grapes, papaya, guava etc. are also cultivated. Grasshoppers are important pests on above crops in plain region of Satara.

Chapter 4
Diversity of Grasshoppers

Introduction

Biodiversity plays an important role in sustainable development of a region or country. Hence, diversity of plant or animal is studied with respect to taxonomy, anatomy, physiology, biochemistry, ethology, genetics, barcoding, etc. In the present study, diversity of grasshoppers was studied with respect to taxonomy of the species.

Taxonomy is the science of classification adapting to previous determined system, with the available catalogue used to providing suitable framework for information retrieval. Classification indicates natural relationship of living things.

Various workers have tried to explain these terms into different manner. According to Mason (1950) taxonomy is the synthesis of all the facts about organisms, in the concept and expression of the interrelationship of organism. Davis and Heywood (1963) define taxonomy as a way of arranging and interpreting information. Heslop- Harrison (1963) says that taxonomy is the study of the principles and practices of classification, in particular the methods, the principles and evens in part the result of biological classification. According to Blackwelder (1967) it is the day to day practice of handling different kinds of organisms. This includes collection and identification of specimen, the publication of data and study of literature.

Taxonomy can be studied as alpha taxonomy, beta taxonomy and gama taxonomy. In alpha taxonomy species are characterized and named, while beta taxonomy used to arrangement of species into natural system of higher and lower categories and gama taxonomy refers to analysis of species variation and evolutionary studies.

Taxonomy is easy tool for understanding the biodiversity and helpful for conservation of species. Taxonomic work affects the activities of Wildlife and its conservation, pest management, environmental problems, nutritional problems, agricultural and forest problems, and many more problems related to environment and human beings.

In present topic taxonomical diversity of grasshoppers was studied. The grasshopper belongs to order Orthoptera.

Order Orthoptera is divided in two suborders, eighteen superfamiles, and fifty three familes. (Hoell *et al.,* 1998).

Classification

Suborder : Ensifera

Superfamily : Grylloidea

Family : Gryllidae

Family : Grylotalpidae

Family : Mogoplistidae

Family : Myrmecophilidae

Superfamily: Hagloidea

Family : Haglidae

Family : Hagloedischiidae

Family : Prophalangopsidae

Family : Tuphillidae

Superfamily : Pasmomimoidea

Family : Phasmomimidae

Superfamily : Rhaphidophoroidea

Family : Rhaphidophoridae

Superfamily : Schizodactyloidea

Family : Scizodactylidae

Superfamily : Stenopelmatoidea

Family : Anostomatidae

Family : Cooloolidae

Family : Gryllacrididae

Family : Stenopelmatidae

Superfamily: Tettigonioidea

Family : HagloTettiigonidae

Family : Tettigoniidae

Suborder : Califera

Infraorder : Acrididae

Superfamily : Acridiodea

Family : Acrididae

Family : Charilaidae

Family : Dericorythidae

Family : Lathiceridae

Family : Lentulidae

Family : Lithidiidae

Family : Ommexechidae

Family : Pamphagidae

Family : Pyrgacrididae

Family : Romaleidae

Family : Tristridae

Superfamily: Eumastacoidea

Family : Chorotypidae

Family : Episactidae

Family : Eumastacidae

Family : Euschmidtiidae

Family : Mastacideidae

Family : Morabidae

Family : Promastacidae

Family : Thericleidae

Superfamily : Locustopsoidea

Family : Araripelocustidae

Family : Bouretidae

Family : Eolocustopsidae

Family : Locustavidae

Family : Locustopsidae

Superfamily : Pneumoroidea

Family : Pnemoridae

Superfamily : Pyrgomorphoidea

Family : Pyrgomorphidae

Superfamily : Tanaoceroidea

Family : Tanaoceridae

Superfamily : Tetrigoidea

Family : Tetrigidae

Superfamily : Trigonopterygoidea

Family : Trigonopterygidae

Family : Xyronoptidae

Infraorder : Tridactylidea

Superfamily : Dzhajloutshelloidea

Family : Dzhajlousthellidae

Superfamily : Regiatoidea

Family : Regiatidae

Superfamily : Tridactyloidea

Family : Cylindrachetidae

Family : Ripipterygidae

Family : Tridactylidae

In past Alexander et al.,(2011), Ananthaselavi *et al.* (2007), Bhowmik (1984, 1990), Chandra *et al.* (2007), Chandra (2007), Chandra and Gupta (2009), Cots (1891), Gupta and Chandra (2009), Hawksworth (1994), Heller *et al.* (2004), Hubbel (1932), Ingrich and Shishodia (1997), Ingrisch and gorai (2001), Ingrisch and murlyrangan (2003), Johanson (1998), Kandibane *et al.* (2004), Karby (1914), Mandal *et al.* (2007), Shishodia (1991, 1995, 1999, 2000, 2006), Shishodia and Gupta (2009), Spungis (2007), Suhil *et al.* (1999), Seetharam *et al.* (2005), Senthilkumar *et al.* (2006), Tandone *et al.* (1976), Uvarov (1925), Willemse and Ingrisch (2004), Hill (2007), Ingrisch (1998), Ingrisch (2002), Jago and Bhowmik (1990), Lookword (2010, 2011), Mass (2009), Priya (2003), Rentz (2009), Smith *et al.* (2004) Song (2006), Steak *et al.* (2007) Willems and Ingrish (2004), Yen (2005). etc contributed on grasshopper diversity from different parts of the world.

Materials and Methods

Adult grasshoppers have been collected from Kolhapur, Satara and Sangli district with the help of insect collecting net at evening and morning. The collected specimen were pinned, dried at 55°C and kept in insect storage box and morphological/ taxonomical studies have been made. The species have been identified by consulting literature of Kirby (1914), Bhowmik (1986) and Mandal *et al.* (2007). The type material time being is with T.V.Sathe, Department of Zoology, Shivaji University, Kolhapur and will be deposited at ZSI, Kolkata.

Morphological Consideration

The body of grasshopper is divisible into head, thorax and abdomen. (Figures 26, 27 and 28)

Head (Figure 29)

Head is short, broad, more or less oval in shape, conical in subfamily Acridinae. In Pyrgomorphidae, margins of vertex are somewhat raised and sharp, fairly broad which form regular depression, called fovolae or tempora. This foveolae is situated on each side of above the antennal socket and between the eyes. Fastigium of vertex present at dorsal side of head in between and beyond the antennae, margin of vertex called lateral carinula. Central surface of head called frontal ridge. It is slightly raised and bounded by a carinula. At lateral side of face, below the eyes present

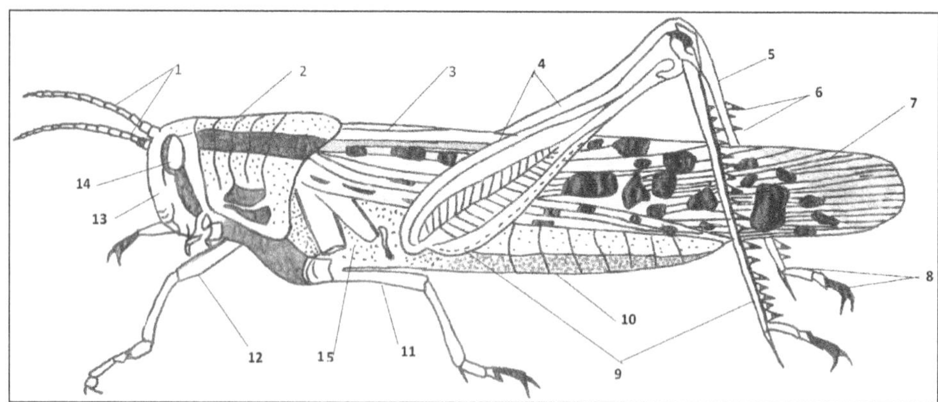

Figure 26: Morphological Features of Short Horned Grasshopper

1: Antennae; 2: Pronotum; 3: Tegmina; 4: Femur of hind leg; 5: Tibia of hind leg; 6: Spines of hind leg; 7: Tip of tegmina; 8: Tarsus; 9: Hind leg; 10: Abdomen; 11: Mid leg; 12: Fore leg; 13: Head; 14: Eye; 15: Thorax.

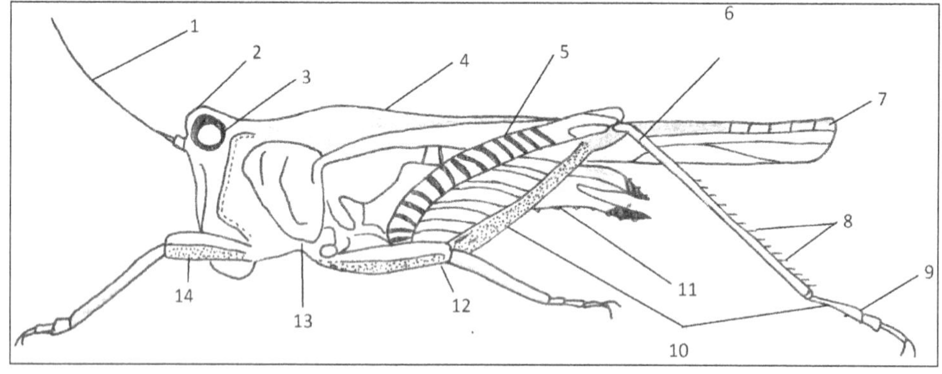

Figure 27: Morphological Features of Grouse Locust

1: Antenna; 2: Head; 3: Eye; 4: Pronotum; 5: Hind femur; 6: Tibia; 7: Wing; 8: Hind tibia; 9: Tarsus; 10: Hind leg; 11: Abdomen; 12: Mid leg; 13: Thorax; 14: Fore leg.

a straight or curved carina called facial carina. Three ocelli present, out of which two are situated at just below the foveolae and remaining one situated at central region of frontal ridge.

Eyes are present on lateral margin of head, they may be rounded, oval, elongated and some time raised above the head.

Antenna (Figure 30) consist of scape, pedicel and flagellum. Scape is generally broad and longer than the pedicel. The pedicel or ring joint, which is more or less globular, remaining joints are flagellum. It is usually

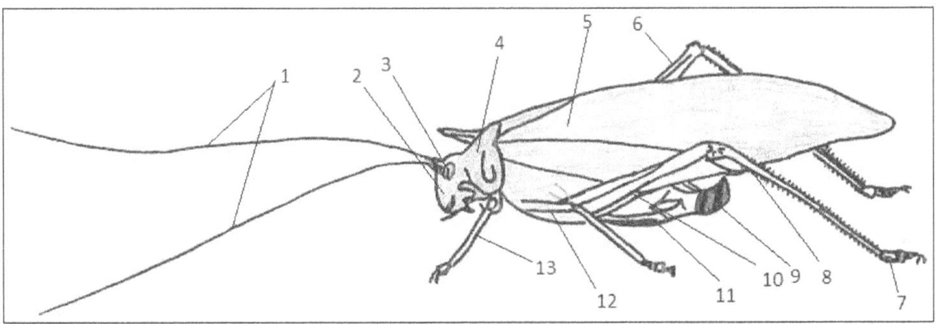

Figure 28: Morphological Features of Long Horned Grasshopper

1: Anntennae; 2: Head; 3: Eye; 4: Pronotum; 5: Tegmina; 6: Hind leg; 7: Tarsus; 8: Tibia; 9: Ovipositor; 10: Hind femora; 11: Abdomen; 12: Mid leg; 13: Fore leg.

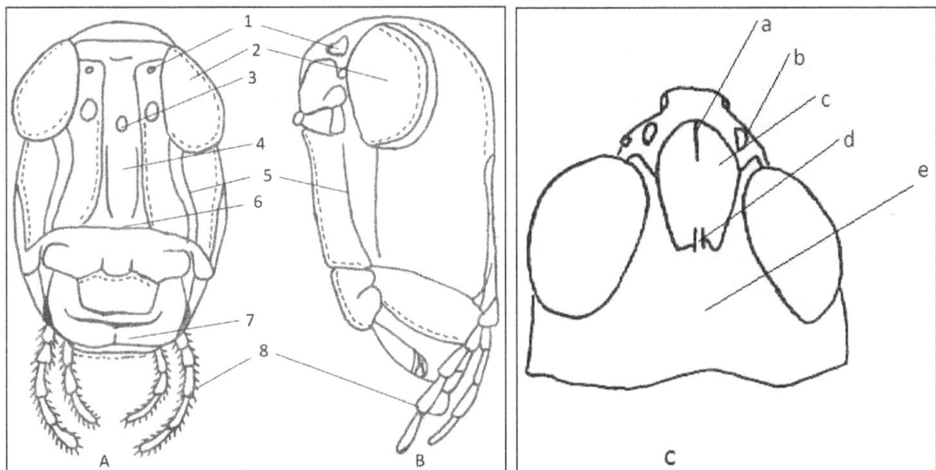

Figure 29: Morphological Features of Grasshopper Head

A: Frontal view; B: Lateral view; C: Dorsal portion of head.

1: Lateral ocelli; 2: Eye; 3: Midle ocelli; 4: Frontal ridge; 5: Lateral carina; 6: Clypeus; 7: Labrum; 8: Maxillae;

a: Fastigial furrow; b: Fastigial foveolae; c: Fastigium of vertex; d: Carinula of vertex; e: Vertex.

filiform (fine round and equally broad) and ensiform type (dorsovetraly flattened, Ist joint beaing wider than their length) present in subfamily Acridinae, Gomphocerinae and family Pyrgomorphidae. The antennal segment varies from 18-30. The antenna is comparatively short, longer than the front femora or longer than body, filiform in Tettigoniidae.

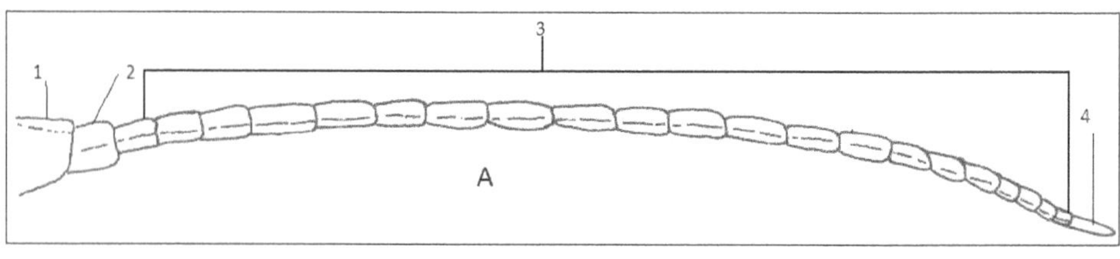

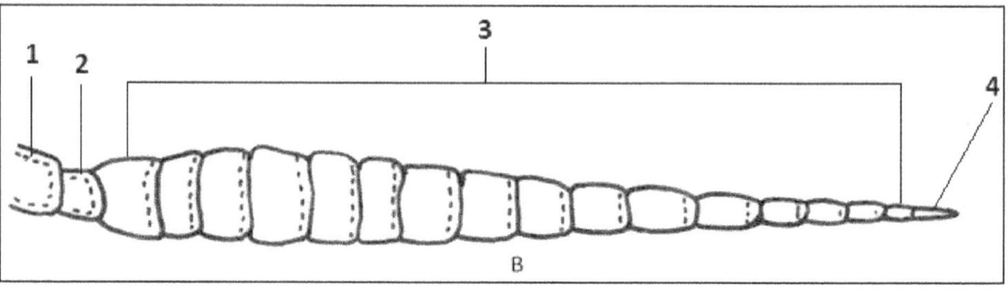

Figure 30: Antennae

A: Simple filiform type; B: Insiform.

1: Scape; 2: Pedicel; 3: Flagellum; 4: Tip.

Thorax

Thorax is divided into three segments namely prothorax, mesothorax, and metathorax. The prothorax is covered by roof shaped plate called pronotum (Figure 31). It is generally broader than the head, anterior margin slightly overlaps on head. It is invariable in shape and size, dorsal side flattened, three carina present, two at lateral side called lateral carina and one at middle called median carina. In some cases lateral and median carina absent. The dorsal side of pronotum is divided by one to three sulcus; the shape, size and positions of sulcus are highly species specific, third sulcus divided to the pronotum in two parts, *i.e.* anterior and posterior, anterior portion termed as prozona and posterior portion termed as metazona. Ventral portion of prothorax called prosternum, which is present in between the base of anterior legs. Mesothoracic and metathoracic segments contain pairs of wings namely fore wings and hind wings. Compact meso and meta sternum, lateral lobes of mesosternum called mesosternal lobes and lateral side of metasternum called metazona, they are separated by metasternal interspace.

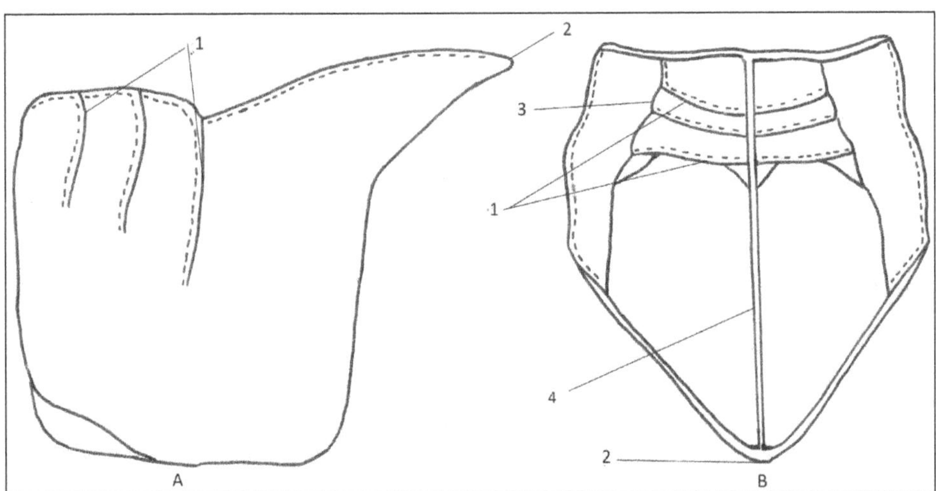

Figure 31: Morphological Features of Pronotum

A: Side view of pronotum; B: Dorsal view of prontum.

1: Sulci; 2: Basal tip of pronotum; 3 : Lateral carina; 4: Median carina.

Fore Wing (Tegmina) (Figure 32)

Elongated, straight, thick, leathery and smaller than the hind wing. Which has large anal area, sub costa long, radius bifurcats more or less, well developed medius and cubitus and reduced anal vein.

Hind Wing (Figure 33)

Fore wing membranous and semitransparent. Anal area of hind wing is much developed, numerous cross veins present. In some species of grasshoppers wings are absent.

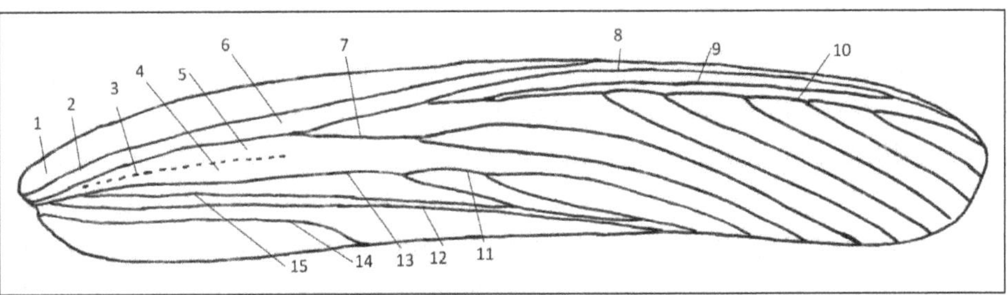

Figure 32: Morphological Features of Tegmina (Fore wing)

1: Mediastinal area; 2: Mediastinal vein; 3: Intercalate; 4 and 5: Anterior and posterior intercalate space; 6: Scapular; 7: Posterior radial; 8: Anterior radial; 9: Middle radial; 10: Branch of radial; 11: Branch of ulnar; 12: Anal; 13: Anterior ulnar; 14: Axillary; 15: Posterior ulnar.

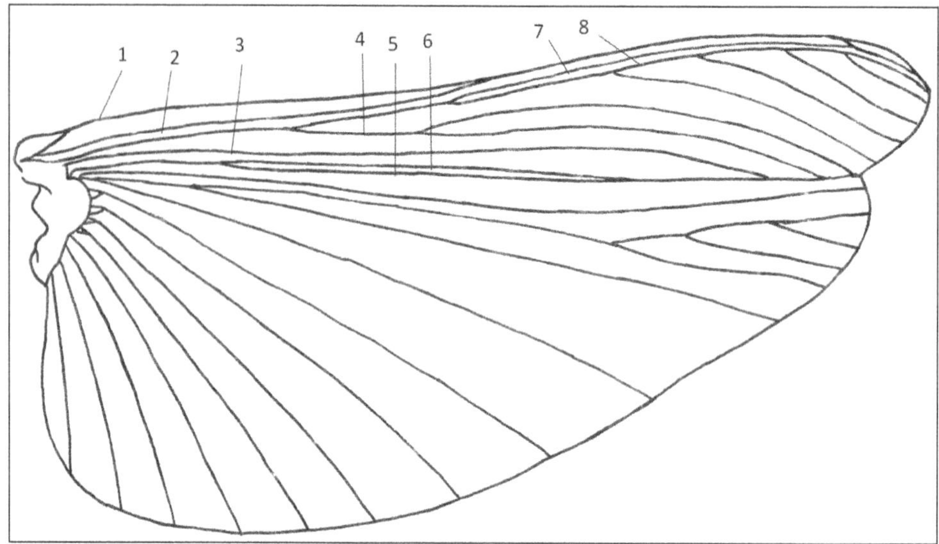

Figure 33: Morphological Feature of Grasshopper Wing (Hind wing)

1: Mediastinal vein; 2.Scapular; 3. Anterior ulnar; 4: Posterior radial; 5: Anal; 6: Posterior ulnar; 7: Middle radial; 8: Branch of radial.

Legs

Legs are well developed. Fore and mid legs are modified for walking and hind legs for jumping. Hind legs (Figure 34) are longer than pro and mid legs, trochanter is very small and fused with coxa; femur very large and stout, swollen at the base and on outer surface present oblique ridges. Hind leg tibia thick with lateral rows of acute spine and two pairs of spines present at extremity called spurs. Tarsus have 3-4 segments.

Abdomen (Figures 35 and 36)

Abdomen consist of 11 segments of which 10 evident and eleventh segment vestigial. 8th sternite of female (Figure 37) and 9th sternite of male (Figure 38) are transformed as sub- genital plate. The 11th tergite is modified as supra anal plate. It is variable structure in males, may be triangular, rectangular or pentagonal, etc. Cercus small, situated above the supra anal plate; subgenital plate in male is simple, navicular with rounded apex and in female ventral surface may be flat or concave with two longitudinal ridges and rounded apex. Ovipositor is on the last segment of female abdomen bearing two upper and lower valves which are often

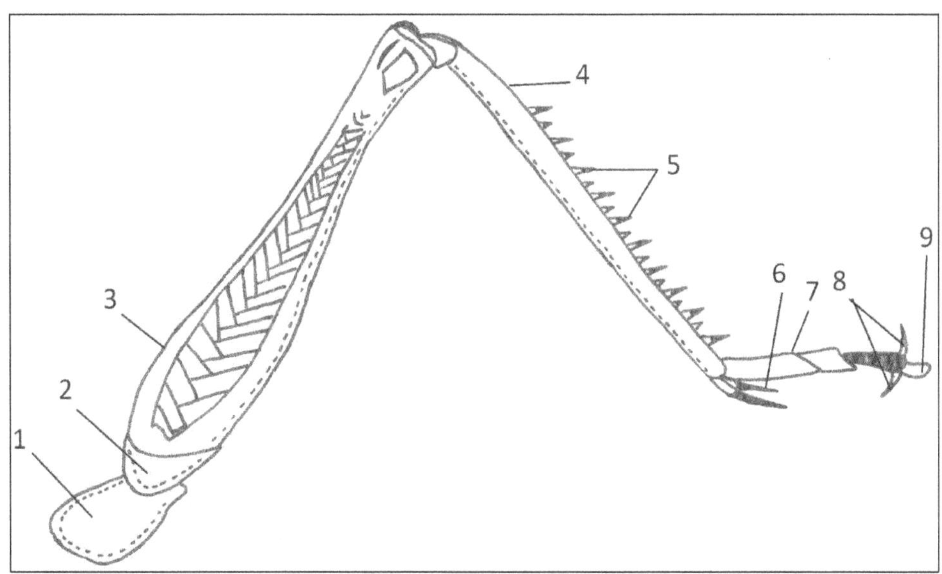

Figure 34: Morphological Features of Hind Leg

1: Coxa; 2: Trochanter; 3: Femur; 4: Tibia; 5: Tibial spines; 6: Calcaria; 7: Tarsus; 8: Claws; 9: Arolium.

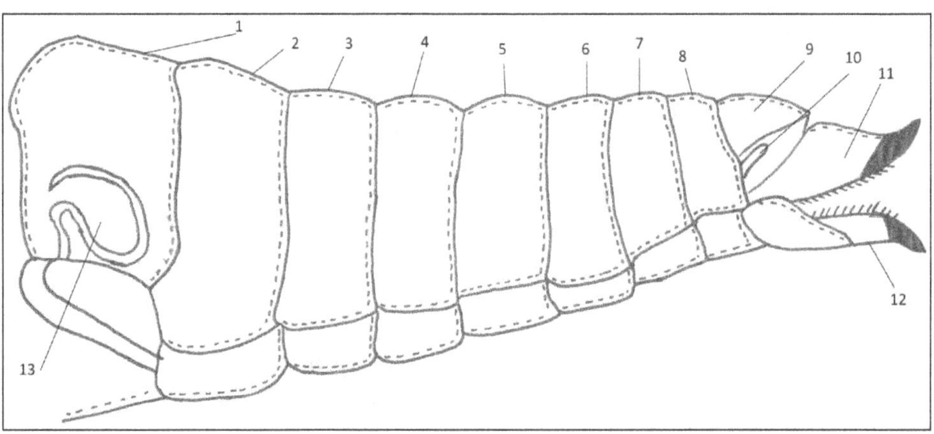

Figure 35: Morphological Features of Short Horned Grasshopper Abdomen

1: Ist abdominal tergite; 2: II nd abdominal tergite; 3: IIIrd abdominal tergite; 4: IVth abdominal tergite; 5: Vth abdominal tergite; 6: VIth abdominal tergite; 7: VIIth abdominal tergite; 8: VIIIth abdominal tergite; 9: Supera annal lamina; 10: cerci; 11: Upper lamellae; 12: Lower lamellae; 13: Auditory oragan.

dentate, it's shape and size highly species specific. In family Tettiigonidae Ovipositor is long, sword shaped or short curved.

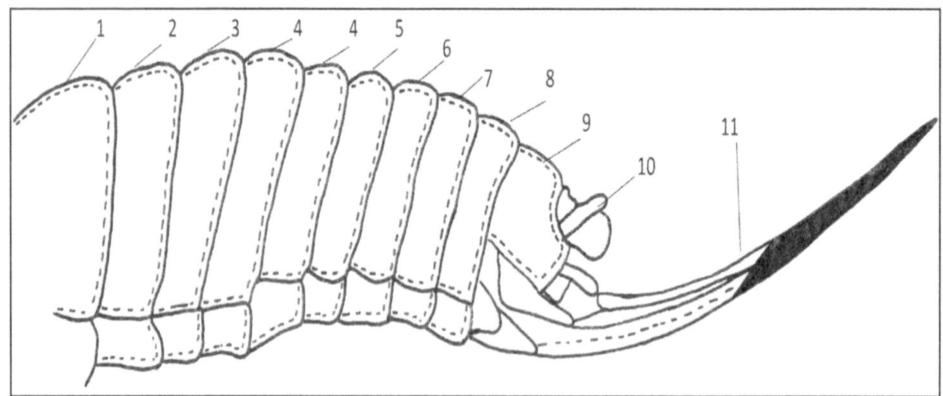

Figure 36: Morphological Features of Long Horned Grasshopper Abdomen

1: Ist abdominal tergite; 2: II nd abdominal tergite; 3: IIIrd abdominal tergite; 4: IVth abdominal tergite; 5: Vth abdominal tergite; 6: VIth abdominal tergite; 7: VIIth abdominal tergite; 8: VIIIth abdominal tergite; 9: supra annal lamina; 10: cerci; 11: ovipositor.

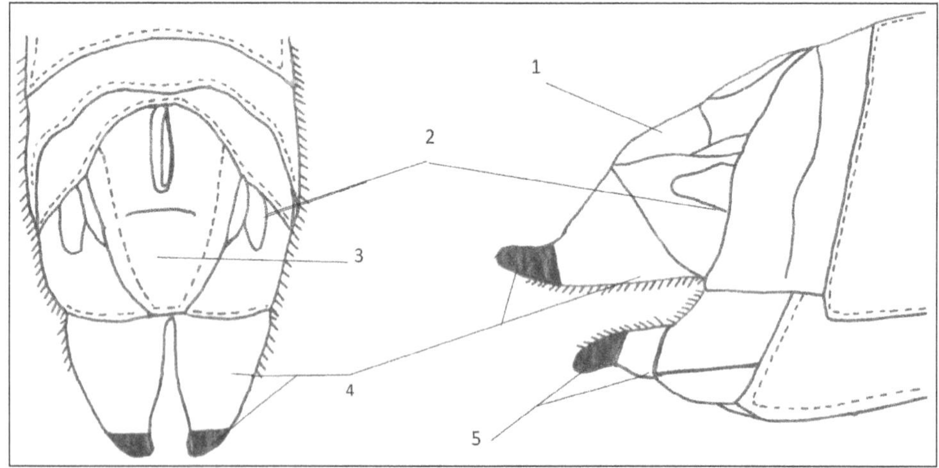

Figure 37: Morphological Features of Female Genitalia of Short Horned Grasshopper

1: Supra annal lamina; 2: Cerci; 3: Supra annal lamina; 4: Upper lamellae; 5: Lower lamellae.

Order Orthoptera is divided in two suborders, eighteen superfamilies, and fifty three families, out of which only four families are selected for studing the grasshopper diversity *viz*. Acrididae, Pyrgomorphidae, Tetrigidae and Tettigoniidae.

Family Acrididae is characterized by,

1. Antennae usually composed of 18 to 30 segments, longer than the front femora but always shorter than the body length.

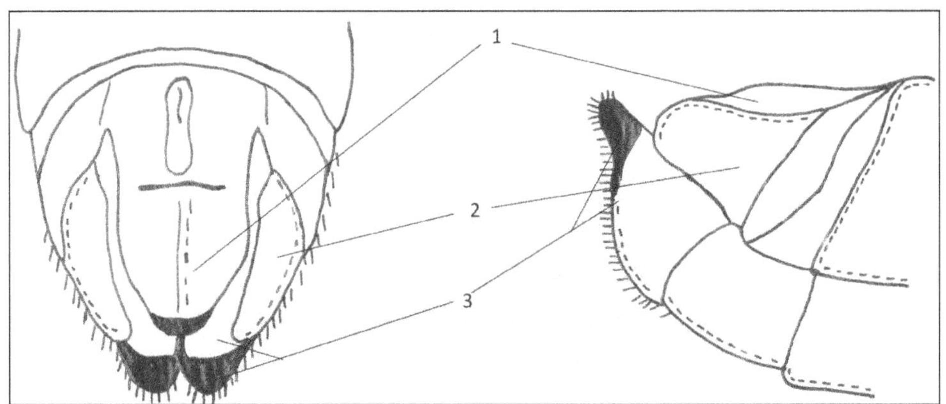

Figure 38: Morphological Features of Male Genitalia of Short Horned Grasshopper 1: Supra annal lamina; 2: Cerci; 3: Subgenital lamina.

2. Pronotum not extend backwards to cover abdomen.
3. Tarsus three segmented with two claws and arolium present at tip.
4. Prosternum armed only in some subfamilies.
5. Pleuro- sternal suture of mesothorax and metathorax subobsolete or obsolete.
6. Foveole of vertex ventro lateral to fastigium.

Family Pyrgomorphidae is closely related to the family Acrididae. It is separated from Acrididae by fallowing characters.

1. Dorsal surface of rostrum nearly horizontal, narrowly separated and medially by vertical fissation of fastigium and forming lateral borders of latter seen from dorsum.
2. Antennae ensiform.
3. Foveolae of vertex relatively large.

Family Tetrigidae is characterized by following characters.

1. Body small, dark brown or grey colour.
2. Antennae short, hair like.
3. Vertex may be broder than the eyes.
4. Pronotum backwardly extend to a hood like covering to the body.
5. Absence of arolium between the tarsal claws.
6. Tegmina highly reduced.

Family Tettiigonidae shows following characters

1. Antennae longer than the body.
2. Stridulatory organs present at cubito anal filed of fore wing.
3. Auditory organs present at the base of fore tibia.
4. Tarsus three to four segmented.
5. Fore wings with pre costal files.
6. Hind wings with more or less well developed.
7. Long saw shaped or short curved ovipositor.

Key to the Families of Order Orthoptera

1. Claws without arolium; pronotum greatly lengthened and extending backward over the entire abdomen, fore wing vestigial and reduced to small scales .... TETRIGIDAE

– Foveolae of the vertex contiguous, superior and forming the extremity of the fastigium; stridulatory mechanism absent .... PYRGOMORPHIDAE

2. Foveolae lateral or inferior, never forming the tip of the fastigium, stridulatory mechanism present .... ACRIDIDAE

– Fore tibia with auditory organ; antennae longer than the body and filiform or thread like................ TETTIIGONIDAE

Key to the Subfamilies of Family Acrididae

1. Prosternal process usually absent; antenna ensiform and body elongate 2

– Prosternal process present; antenna and body different 4

2. Stridulatory mechanism present, in form of a series of peg like hair, along inner lower side of posterior femora .... Gomphocerinae

– Stridulatory mechanism not presents on inner side of posterior femora. Fastigium little delivent; eyes generally longer than the infraocular area 3

3. Head with high acute profile; face retreating and angulate; tegmina without intercalary vein .... Acridinae

- Head with rounded profile; face almost vertical, rarely oblique; tegmina with an intercalary vein in medial area .... Oedepodinae

4. Radial area of tegmina with a series of regular, parallel sound producing veinlets .... Hemiacridinae

- Sound producing veilnets radial area of tegmina absent, if apterous, then tympanum present (at least reduced) 6

5. Lower external lobe of hind leg posterior knee with spine like apex .... Oxynae

- Lower external lobe of hind leg knee with rounded apex, sub acute or angular,but not spine like 7

6. Mesosternal interapace closed; last abdominal tergaite in male without well developed furcula .... Tropidopolinae

- Mesosternal interspace open, last abdominal tergaite in male well developed furcula 8

7. Mesosternal lobes acute anguler or obtus angular or rounded, not triangular 9

- Mesosternal lobes rectangular .... Cyrtacanthacridinae

8. Dorsum of pronotum weakly tectiform or flat, with median and lateral carina linear (lateral carinasum time oblited); male cercus highly compressed, lobioform or subacute down .... Eyprepocnemidinae

- Dorsum of pronotum of different shape; lateral carina if present, not linear; male cercus variable, but not with strongly compressed, lobioform or subacute .... Catantopinae

Key to Genera of Subfamily Acridinae

1. Apex of tegmina acute or sub acute; "knee" lobes of posterior femur acute- pointed; head elongated; prosternal process absent; medial area of wing widened, forming; tegmina with dense reticulation .... *Acrida* (Linnaeus)

- Apex of tegmina round or obtuse; "knee" lobes of posterior femur with rounded apex; head small; antenna ensiform, at least on basal half; lateral carina of pronotum linear, contnus; fastigium of vertex concave and often with a strong median dividing carina; posterior femora thickned at bases .... *Phlaeoba* Stal.

2. Size large; head conical, ascending; fastigium broad, laminate and struncate at apex .... *Truxalis* Fab.

Key to the Species of Genus *Acrida* Linnaeus

1. Tegmina without distinct coloured mrgin; length of body 50 mm .... *exaltata* (Walker)

- Tegmina very narrowly edged with pinkish brown; length of body 30mm .... *liguberis* Burr.

Key to the Species of Genus *Plaeoba* Stal

1. Antennae unicodolus; fastgium of vartex above with a continuous median carina .... *infumata* Brunner

- Antennae narrowly ensiform at base; fastigium of vartex broad, shallowly concave, with strong median carina .... *panteli* Boliver

Key to the Genera of Subfamily Gomphocerenae

1. Fastigium litel longer than broad; antennae narrowly ensiform; tegmina some what acute at apex; posterior femora reaching upto abdomen .... *Gelastorrhinus*, Brunn.

- Fastigium of vertex usually not carinated, antennae distinctly ensiform .... *Ischnacrida* Stal.

Key to the Species of Genus *Gelastorrhincus,* Brunn.

1. Antennae ensiform, long flattend from the 4th joint to the 12th tegmina longer than the abdomen .... *laticoris* Serv.

- Antennae ensiform as long as head and pronotum together, antennae, head and pronotum rufose brown above .... *semipictus* Walk.

Key to the Species of Genus *Ischnacrida* Stal.

1. Wings stained with pink towards the base; hind femora with a continuous silvery longitudinal strip on the median outer area .... *convergence* Walk.

– Wings not stained with pink forward the base; hind femora yellowish green, genicular lobes marked with black within and below .... *tarsalis* Walk.

Key to the Genera of Subfamily Oedepodinae

1. Internal spurs of posterior tibia equal, normal; posterior margin of matozona rectangular with tip rounded off 2

– Internal spur of posterior tibia greatly unequal lower one much longer than the other, abruptly hooked at apex, very acute .... *Heteropternis* Stal.

2. Head and pronotum granulose, latter bituberculate medially in front; intercalary vein weakly serrate .... *Dittopternis* Saussure

– Head and pronotum granulated; posterior margin of metazoan obtuse angular; intercalory vein strongly serrated and extending to distal apex of medial area .... *Aiolopus* Faber

3. Pronotal x- marking with anterior and posterior arms continuous; posterior arms generally curved and with convergent apices; posterior margin of metazona rectangular, never rounded .... *Gastrimargus* Saussure

– Pronotum with well marked crest; median carina of pronotum strongly raised in protozona forming two tooth like projections .... *Trilophidia* Stal.

4. Pronotum without distinct median carina valves of ovipositor covered by the supra anal laminae .... *Dociostaurus* Walk.

– Pronotum with lateral carina sharply angulated inward before the middle .... *Stauroderus* Bol.

5. Wings often coloured at the base, with central black band .... *Peternoscirta* Sau.

Key to the Species of Genus *Heteropternis* Stal.

1. Pronotum velvety, paler or darker above; antennae brown paler at base ciostarsus, Hind femora yellowish, irregular spotted .... *respondence* Walk.

– Pronotum paler behind; antennae reddish; hind femora brown outside, lined with grey .... *pertia* Walk.

Key to the Species of Genus *Detopternis* Sussure

2. Head slightly granular; Pronotum with the principal sulcus slightly before the middle .... *ceylonica* Sauss.

– Head and pronotum strongly granular 2

3. Pronotum rugose, granulated and strogly carinated, lateral lobes rectangular behind .... *venusta* walker

– Pronotum with the tip not very acute, the margin slightly waved, lateral lobes oblique and rounded behind .... *zebrata* Sauss.

Key to the Species of the Genus *Ailopus* Fiber

1. Fastigium of vertex with forward angle more acute; frontal ridge flat gradually narrowing; apical half tibia red .... *tamulus thalasinus* Fab.

– Fastigium of vertex with forward angle, not more acute; apical half of hind tibia blue .... *offince* Bol.

Key to the Species of Genus *Gastrimargus* Saussure

4. Fastigium of vertex concave with median carina continuating over vertex, pronotum moderately tectiform .... *africanus africanus* Saussure

– Fastigium of vertex not concave; pronotum acutely angulated behind, with a strong pale median carina .... *transversus* Thunb.

Key to the Species of Genus *Trilophidia* Thunberg

1. Pronotum rugose with a high median carina, forming two teeth in front; wings not yellow at base .... *annulata* (Thunberg)

– Pronotum rectangular behind; wings yellow at base ....*cristella* Stal.

Key to the Species of Genus *Dociostaurus* Fieb.

1. Pronotum with distinct median stripe; tegmina with spots on radial area .... *decisus* Walk.
- Pronotum without distinct median strip; tegmina with some small spots on radial area .... *mundus* Walk.

Key to the Species of Genus Stauradorus Bol.

2. Slight median carina on the head and pronotum, which are lined with black .... *bicolar* Charp.
- Slite median carina on the head and pronotum, which are lined with black .... *indica* sp nov.

Key to the Species of Genus Pternoscrita

1. Wings red at base; hind femora buff or bluish grey outside, with black spots .... *cinctifemur* Walk.
- Wings yellow at base; hind femora yellowish banded with brown .... *caliginosa* De hann
2. Wings yellow towards tip and reddish towards the base, hind femora stout yellowish, with brown patches behind the extremity .... *sahyadricus* sp nov.

Key to Genera of Subfamily Hemiacridane

1. Male cercus large with apex bifurcated, male supra anal plate narrow,long, with sides excurved towards the apex .... *Hieroglyphus* Krauss
- Male cercus very large, trilobite, male supra anal plate broader than long, with straight side narrowing to acute angular apex .... *Parahiroglypus* Carl.
2. Male cercus small, not bifurcated at apex; male supra anal not as above, Pronotum distinctly tricarinate; fastigium of vertex not more produced before the eyes, prosternal tubercle spathulate .... *Spathosternum* Kraus

Key to the Species of Genus *Hieroglyphus* Kraus

1. Antennae unicoloured, basal joint yellowish green, tipped with yellow .... *banian* (Fab.)
- Antennae with yellow-brown band, basal joint green, tiped with brown .... *kolhapurensis* sp nov.

Key to the Species of Genus *Spathosternum* Krauss

1. Prosternal band well marked; male curcus small not bifurcated at apex; tegmina light brown towards the base and subhiline beyond .... *prasinifrum prasiniferum* Walker
- Postocular band obsulate, or suffused green; male curcus bifurcated at apex; tegmina with a short brown strip on the disk, marked with whitish veins before and behind .... *venulosum* Stal.

Key to Species and Subspecies of Genus *Oxya* Servile

1. Male cercus conical or if compressed then narrowing towards apex, which is obtuse or truncate

   Female subgenital plate, on ventrum, markedly concave, bordered posteriorly by two longitudinal, toothed lateral ridges .... *hyla hyla* Serville
- Male cercus with subacute or truncate apex, supra anal plate with basilateral fold

   Female posterior ventral basivalvular sclerites having one or two tooth like spine on its inner ventral margin; lateral longitudinal ridges of ovipositor with two teeth at apices .... *japonica japonica* Blanchard

Key to the Genera of Subfamily Cyrtacanthacridinae

1. Conical compressed apex, straight prosternal tubercle, vertical or slightly inclined backwards in the direction of mesosternum; Pronotum with high median carina in both sexes, anterior part comb shaped; in male sub genital plate trilobat, with 2 rounded incisons at apex .... *Anacridium* Uvarov

- Acute or sub-acute apex backwardly curved prosternal tubercle touching almost mesosternum, inflatened in middle Tegmina with straight venation in apical part, transverse vein forming almost right angle with principle veins; prosternal process almost cylindrical with rounded apex .... *Patanga* Uvarov

2. Pronotum little constricted; integument fairly rugose or dotted; male cercus with subacute apex .....*Cyrtacanthacris* Walker

Key to Species of Genus *Patanga* Uvarow

1. Pronotum stout, hind margin obtus to triangular; wing base rosy violet or colorless; male genital plate long, curved upwords, conical and apex pointed .... *succincta* (Johansson)

- Prosternum distinctly compressed latterly; wing base not rosy; tegmina with coarse and irregular reticulation .... *japonica* (Bol.)

Key to the Species of Genus *Cyrtacanthacris* Walker

1. Pronotum with velvety blackish brown band on dorsolateral side, tegmina with transvers oblique, irregular spots .... *tatarica* (Linnaeus)

- Pronotum strongly rugose, with the median carina forming a strong ridge; tegmina green, rather broad, roundly truncate at the tips .... *rosea* De Green

Key to the Genera of Subfamily Catantopinae

1. Pronotum subcylindrical, slightly narrowing Prosternal tubercle thick, cylindrical or slightly anterior-posteriorly compressed with rounded apex ....*Catantops* Schaum.

- Pronotum not subcylindrical. Prosternal tubercle never with rounded apex 2

2. Pronotum constricted. Prosternal tubercle conical .... *Xenocatantopes* Dirh and Uvarov

- Pronotum flattened, porsternal tubercle laterally compressed .... *Stenocatantopes* Uvarow

Key to the Species of Genus *Catantops* Schaum

1. Frontal ridge parallel side; impressed in middle, starting from median ocellus up to clypeus; fore wing longer than the abdomen, finely mottled with brown, radial area pale .... *pinguis* Walker

– Frontal ridge very slightly depressed in male; fore wing subhyline, brown towards base, and motteled with brown, posterior area pale .... *erubescens* (Walker)

Key to the Genera of Subfamily Eprepocenamidae

1. Antenna filiform. Frontal ridge slightly constricted at apex. Male circus long, thick, with coriaceous apical portion; subgenital plate conical, compressed towards apical region, and almost pointed at apex .... *Choroedocus* Bolivar

– Male cercus narrow, may be slightly compressed, but always acute or subacute apical 2

2. Hind femur moderately long moderately produced beyond abdomen, not inflated basally, not strongly narrowing on apical half. Posterior tibia with sparse spine. Prosternal process almost spathulate, cylindrical, with rounded apex .... *Eyprepocenamis* Fieb.

3. Hind femur long, produced far beyond end of abdomen, inflated basally and strongly narrowed on apical half. Posterior tibia densely spined. Prosternal process almost spathulate, with rounded sometimes slightly inflated apex .... *Tylotropidius* Stal.

Key to the Species of Genus *Eyprepocenamis* Fieber

1. Tegmina with numerous brown spots; posterior femora with a longitudinal black strip on the outer side; tibia and tarsus brown .... *alacris alacris* Serv.

– Tegmina with numeorus green or brown spots in female, anal area green; posterior femora with no black stripe on the outer side; tibia brown base, with pale ring .... *pulchra* Bol.

Family–Acrididae

Subfamily–Truxalinae

Tribe–Truxalini

Truxalis indica **Bolivar, 1902 (Plate 7, Figures 47 to 54)**

Green; head conical, longer than pronotum; eyes oval; tegmina and pronotum without black lines, body and tegmina (Figure 52) with pinkish markings, tegmina green narrow towards the end, slightly longer than the wings; wings hyaline, pinkish at base; a row of small pointed backwardly directed spines present on lateral side of tibia. The tibia (Figure 54) mostly green; head 13 mm long; antennae (Figure 51) 16.8 mm, scape 1.8 mm long and 1.3 mm wide, pedicel 1 mm long and 0.9 mm wide, flagella 14 mm long; pronotum 9 mm long; tegmina 41 mm long; abdomen 30 mm long; body length 58 mm long. Mostly feed on paddy, wheat and other different grasses.

Flagellar formula: 1L/W = 1.32, 4L/W = 1.58, 7L/W = 2.5, A = 1.8

Family–Acrididae

Subfamily–Acridinae

Tribe–Acridini

Acrida exaltata **Walker, 1859 (Plate 6 Figures 39 to 40)**

A. exaltata male green, female not perfect green; a brown coloured sculpturing present on thorax and tegmina. Head elongated, longer than pronotum; eyes oval, brownish coloured; yellowish band present on lateral side of meso and meta thorax; femur (Figure 45) acute, pointed; tegmina without distinct coloured margin, apex of tegmina (Figure 44) acute and long. Dorsal side of tegmina is brown, laterally show zigzag markings. Knee lobes of hind femora posterior; tibia green. A row of backwardly directed spines present on lateral side of tibia. In female, head 17 mm long; antenna (Figure 43) 13.9 mm, scape 1.5 mm long and 1.2 mm wide, pedicel 0.9 mm long and 0.7 mm wide, flagella 13.9 mm long; pronotum (Figures 41,42) 9 mm long; meso and meta thorax 10 mm long; abdomen 29 mm long; tegmina 52 mm long; body length 58 mm long; wing span 100 mm; in male, head 12 mm long; pronotum 6 mm long; meso and meta thorax 6 mm long; abdomen 25 mm long; tegmina 34 mm long; body

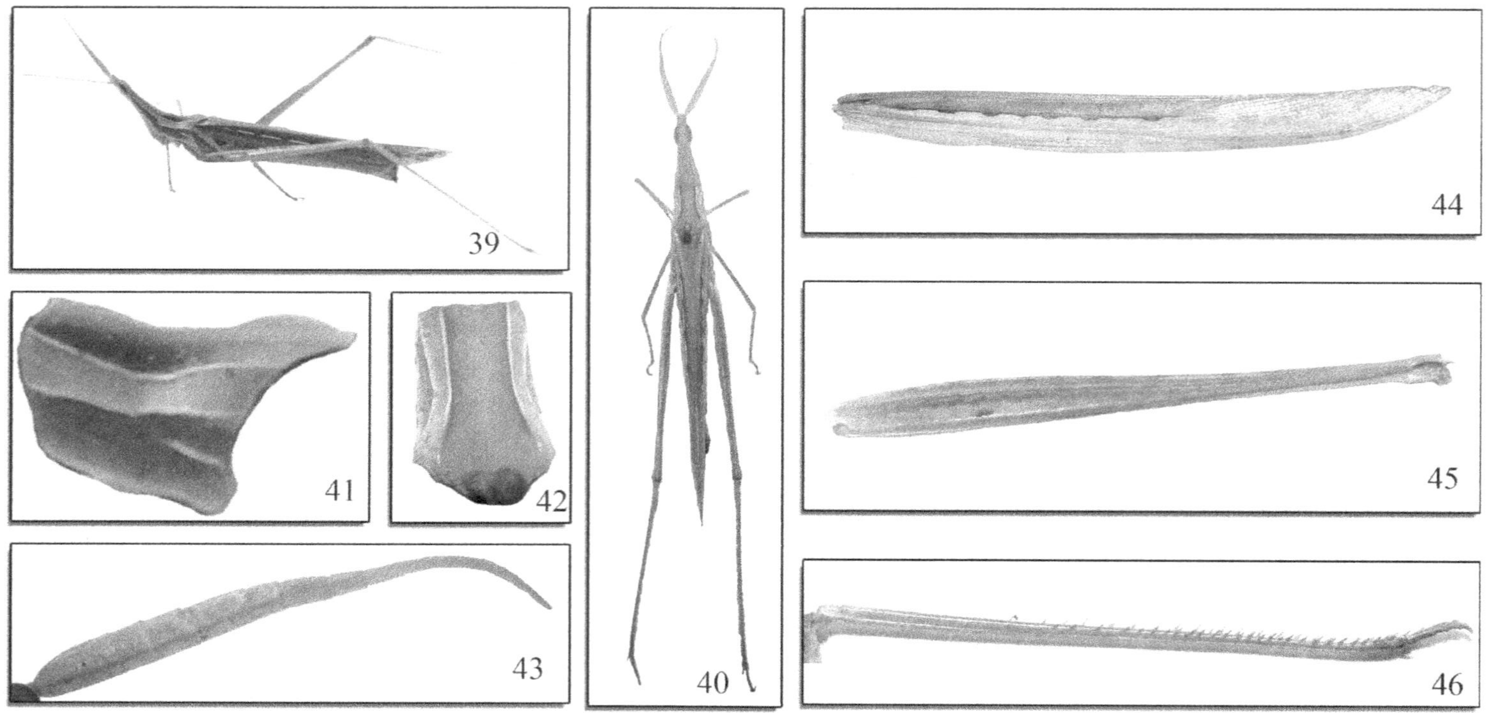

Plate 6: Figure 39: *Acrida exaltata* (Walk); Figure 40: *A. exaltata* dorsal view; Figure 41: Pronotum lateral view; Figure 42: Pronotum dorsal view; Figure 43: Antenna; Figure 44: Tegmina; Figure 45: Femur; Figure 46: Tibia

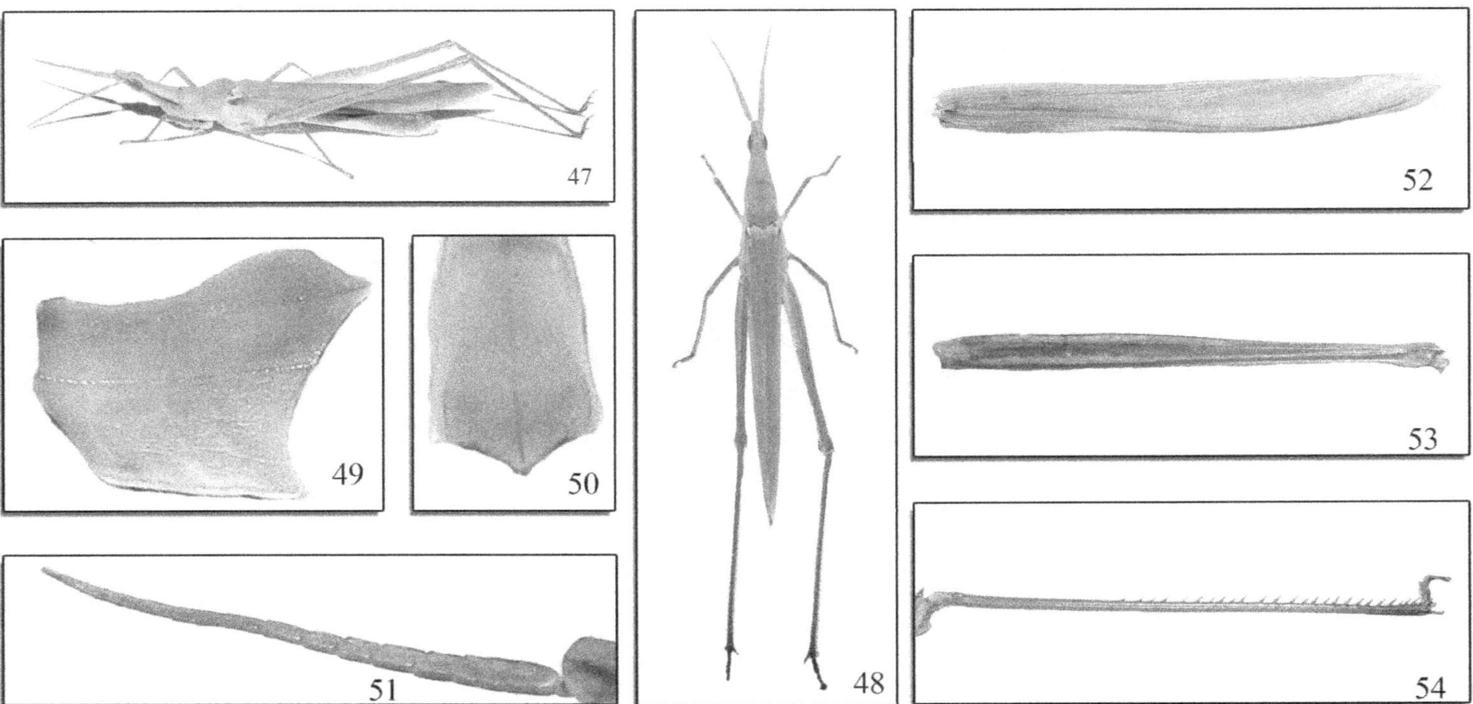

Plate 7: Figure 47: *Truxalis indica* (Bolivar), female; Figure 48: *T. indica* dorsal view; Figure 49: Pronotum lateral view; Figure 50: Pronotum dorsal view; Figure 51: Antenna; Figure 52: Tegmina; Figure 53: Femur; Figure 54: Tibia

length 46 mm long. They feed on maize, paddy, wheat, jowar and vegetables like tomato, brinjal.

Flagellar formula: 1L/W = 1.32, 4L/W = 1.19, 7L/W = 1.66, A = 1.39

Family–Acrididae

Subfamily–Acridinae

Tribe–Phlaeobini

***Phlaeoba infumata* Brunner, 1893 (Plate 8, Figures 55 to 63)**

Body grey, brown; head small; antennae (Figure 58) unicodolus, ensiform, at least on basal half; lateral carina of pronotum (Figures 59, 60) linear, continuous; fastigium of vertex shallowly concave and often with a median dividing carina; eyes oval, brown; median carina continuous from vertex to tip of pronotum; legs brown grey, at anteroventral side of fore and middlegs tibia present a row of small pointed spines; posterior femora (Figure 62) thickened at the bases, on outer carina present a black spot; tibia greyish, a black tipped 10-11 spines present on lateral side of hind tibia; tarsus reddish, tegmina (Figure 61) longer than the abdomen. In female; head 8 mm long; antennae 12.5 mm, scape 2 mm long and 1.7 mm wide, pedicel 1 mm long, flagellum 9.5 mm long; pronotum 5 mm long; meso and meta thorax 7 mm long; tegmina 25 mm long; abdomen 15 mm long; hind femora 17 mm long; hind tibia (Figure 63) 14 mm long; wing span 52 mm long. In male, antennae 9 mm long; head 5.5 mm long; pronotum 4 mm long; meso and meta thorax 4.5 mm long; abdomen 10 mm long; tegmina 18 mm long; wing span 38 mm long; hind femora 13 mm long; hind tibia 11 mm long. This species feed on young sugarcane, maize, wheat, jowar and grasses.

Flagellar formula: 1L/W = 1.25, 4L/W = 1.15, 7L/W = 1.3, A = 1.2

Family–Acrididae

Subfamily–Acridinae

Tribe–Phlaeobini

***Phlaeoba panteli* Bolivar, 1902 (Plate 9, Figures 64 to 70))**

Brownish, in female yellow band running from the tip of head to tip of tegmina (Figure 68), antennae flat ensiform; pronotum (Figure 67) smooth, lateral carina bended middle; eyes oval; tegmina upto the tip of

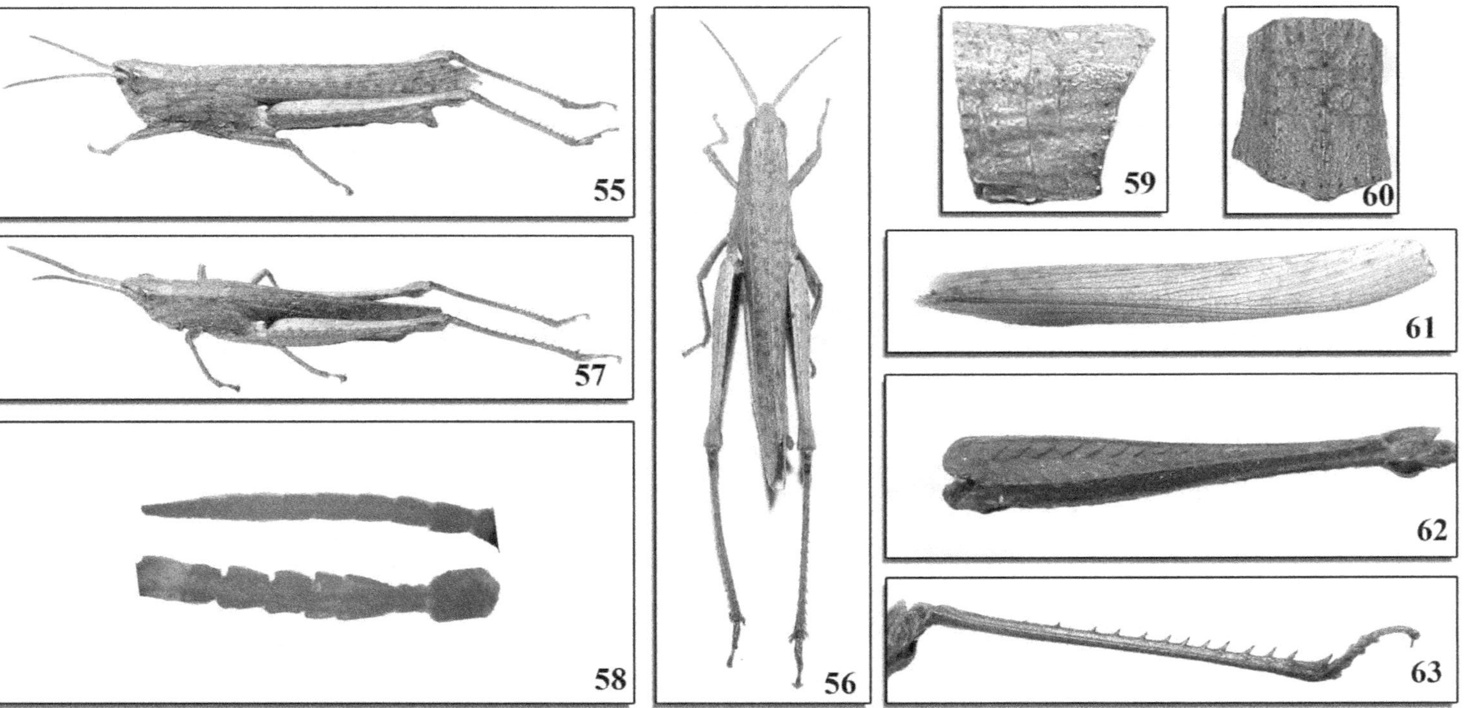

Plate 8: Figure 55: *Phlaeoba infumata* Brunn (Female); Figure 56: *P. infumata* dorsal view; Figure 57: *P. infumata* (male); Figure 58: Antenna; Figure 59: Pronotum (lateral view); Figure 60: Pronotum (dorsal view); Figure 61: Tegmina; Figure 62: Femur; Figure 63: Tibia

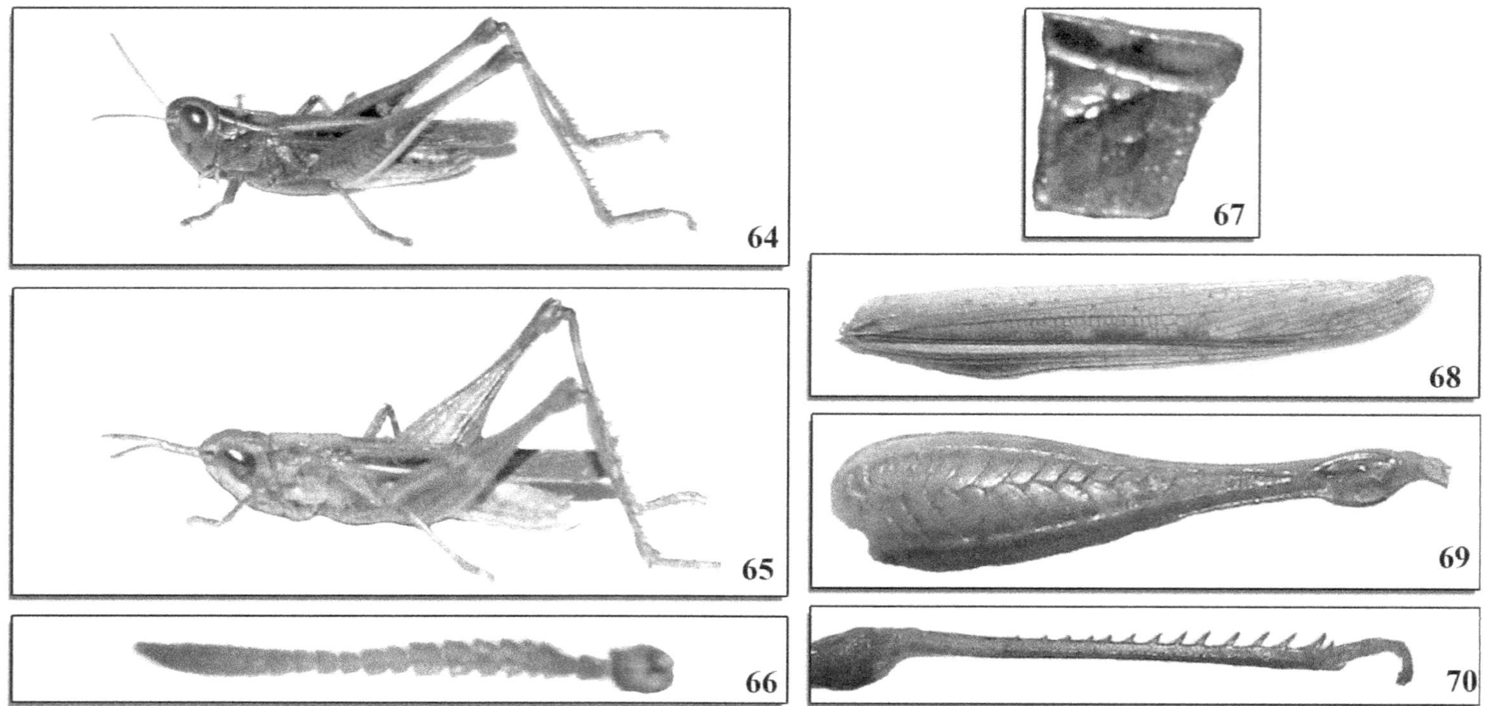

Plate 9: Figure 64: *Phlaeoba panteli* (Bol.) (male); Figure 65: *P. panteli* (Female); Figure 66: Antenna; Figure 67: Pronotum (Lateral view); Figure 68: Tegmina; Figure 69: Femur; Figure 70: Tarsus

abdomen; legs grey, brown; abdomen also brown; hind femora (Figure 69) brown, stout at the base; tibia (Figure 70) also brown, 9-10 black tipped spine present on lateral side of tibia. In female, antennae (Figure 66) 10.8 mm long, scape 1.2 mm long and 1 mm wide, pedicel 0.6 mm long and 0.5 mm wide, flagella 9 mm long; head 5 mm long; pronotum 4 mm long; tegmina 13 mm long; wing span 27 mm; abdomen 11 mm long; hind femora 12 mm long; hind tibia 9 mm long; in male, head 4.5 mm long; antennae 5 mm long; pronotum 3.5 mm long; meso and meta thorax 4 mm long; tegmina 13 mm long; abdomen 9 mm long; wing span 27 mm long.

Flagellar formula: 1L/W = 0.62, 4L/W = 0.44, 7L/W = 0.66, A = 0.57

Family–Acrididae

Subfamily–Gomphocerinae

***Gelastorrhinus laticornis* (Serville, 1839) (Plate 10, Figures 71 to 79)**

Body green; antennae (Figure 71) reddish green; eyes oval, brown; lateral carina of pronotum yellowish brown, median carina slightly pointed; tegmina (Figure 77) pointed at the tip and longer than the wings; hind femora (Figure 78) green shorter than the abdomen; tibia (Figure 79) also green with backwardly directed 16-18 spines. In female head 9 mm long; antennae 20 mm, scape 3 mm long and 2 mm wide, pedicel 1.1 mm long and 1.8 mm wide, flagella 16 mm long; prothorax 4.5 mm long; pronotum 7mm long, meso and meta thoracic segment 9 mm long; abdomen 24 mm long; tegmina 35 mm long; wing span 7 mm. In male head 7 mm long; prothorax 3 mm long; meso and meta thorax 6 mm long; abdomen 15 mm long; body length 31 mm; tegmina 27 mm long; wing span 50 mm.

It is a pest of maize, lady finger, wheat, paddy, etc.

Flagellar formula: 1L/W = 1, 4L/W = 0.86, 7L/W = 1.58, A = 1.44

Family–Acrididae

Subfamily–Gomphocerinae

***Gelastorrhinus semipictus* (Walker, 1870) (Plate 11, Figures 80 to 87)**

Body greenish brown; eyes oval; antennae (Figure 82) brown; head conical; pronotum tricarinate, a brown band runs over the head and

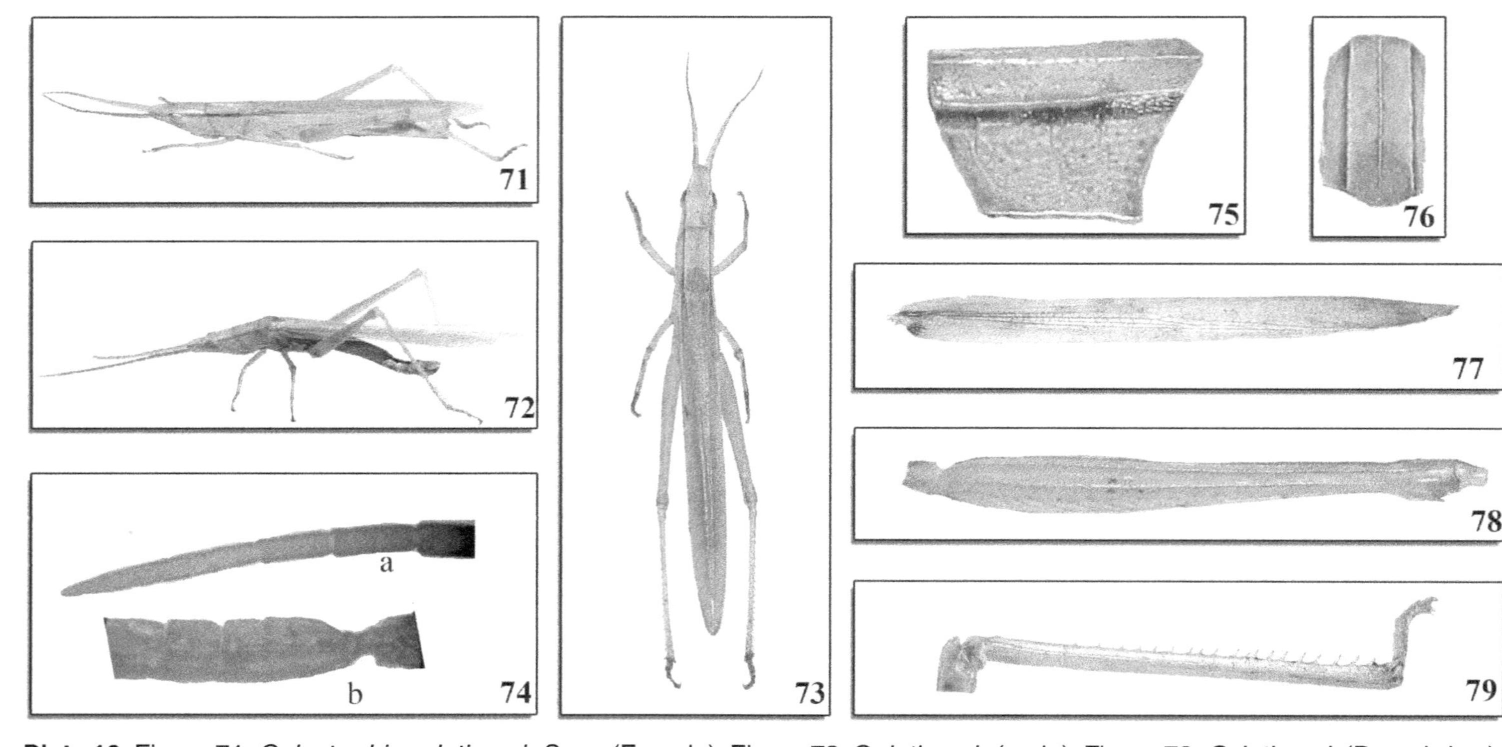

Plate 10: Figure 71: *Gelastorrhinus laticornis* Serv. (Female); Figure 72: G. *laticornis* (male); Figure 73: *G. laticornh* (Dorsal view); Figure 74: Antenna (a–Anterior region, b–Posterior region); Figure 75: Pronotun (Lateral view); Figure 76: Pronotum (Dorsal view); Figure 77: Tegmina; Figure 78: Femur; Figure 79: Tibia

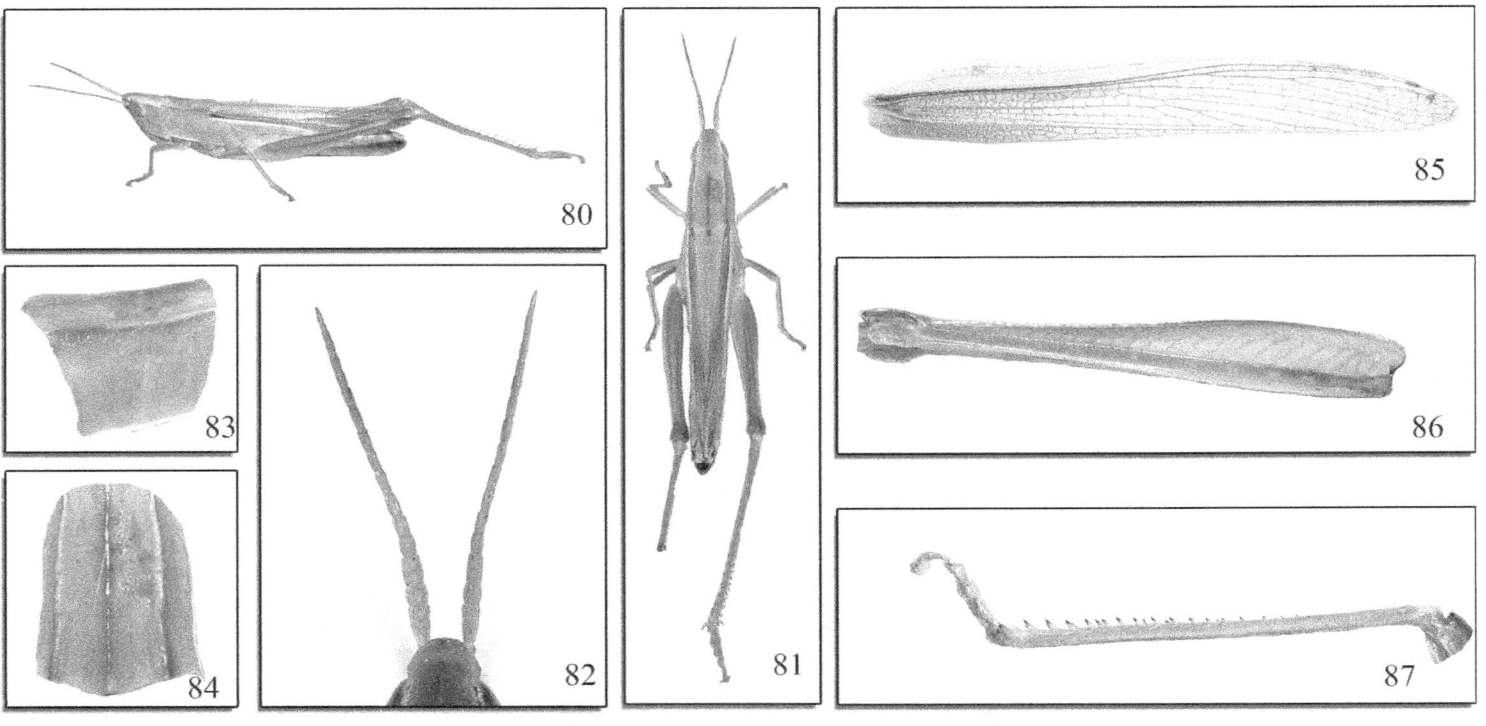

Plate 11: Figure 80: *Gelastorrhinus semipictus* Walk. (Female); Figure 81: *G. semipictus* (Dorsal view); Figure 82: Antennae; Figure 83: Pronotum lateral view; Figure 84: Pronotum dorsal view; Figure 85: Tegmina; Figure 86: Femur; Figure 87: Tibia

pronotum, lateral carina slightly divergent by two sulci, three sulci present on the pronotum, first short and remaining two long up to ventral region of pronotum; lateral side of head and pronotum (Figure 83,84) yellowish green; tegmina (Figure 85) pointed at the tip, hyaline, yellowish green at costa; dorsal side brown mid and fore legs brown, hind femora (Figure 86) green towards the base and brownish towards the tip; hind tibia (Figure 87) pinkish red with 17 black tipped spines, first tarsal segment thick, second short and third narrow with claws. In female, antennae 8 mm long; head 12 mm long; pronotum 8 mm long; thorax 11 mm long; abdomen 22 mm long; tegmina 30 mm long; wing span 54 mm.

Flagellar formula: 1L/W = 1, 4L/W = 0.86, 7L/W = 1.58, A = 1.14

Family–Acrididae

Subfamily–Gomphocerinae

Ischnacrida convergens **(Walker, 1902) (Plate 12, Figures 88 to 95)**

Body yellowish,cylindrical; antennae (Figure 93) flat, pointed at tip; vertex pointed, frontal ridge laterally compressed in between the antennae, frontal ocelli white; front region of head brown; eyes round, silvery white band runs from base of antennae to the base of hind coxa on ventrolateral side of head and thorax; small whitish longitudinal line present on pronotum; fore and mid leg short, hind leg long; longitudinal faint silvery band present on outer side of hind femora (Figure 95) and on inner side black line with white spot present, also black spots present at the knee on inner side of hind femora; tibia (Figure 96) yellowish, lateral side of tibia present a row of black tipped spines. In female; head 8 mm long; antennae 16 mm, scape 1.7 mm long and 1.8 mm wide, pedicel 1.3 mm long and 1.8 mm wide, flagella 13 mm long; pronotum 7 mm; meso and meta thorax 10 mm long; abdomen 33 mm long; tegmina 37 mm long; hind leg femora 19 mm long; hind leg tibia 17 mm long. In male antennae 12 mm long; head 7 mm long; pronotum 5.5 mm; meso and meta thorax 7 mm long; tegmina 28 mm long; abdomen 35 mm long; wing span 60 mm.

Flagellar formula: 1L/W = 0.48, 4L/W = 0.38, 7L/W = 0.58, A = 0.48

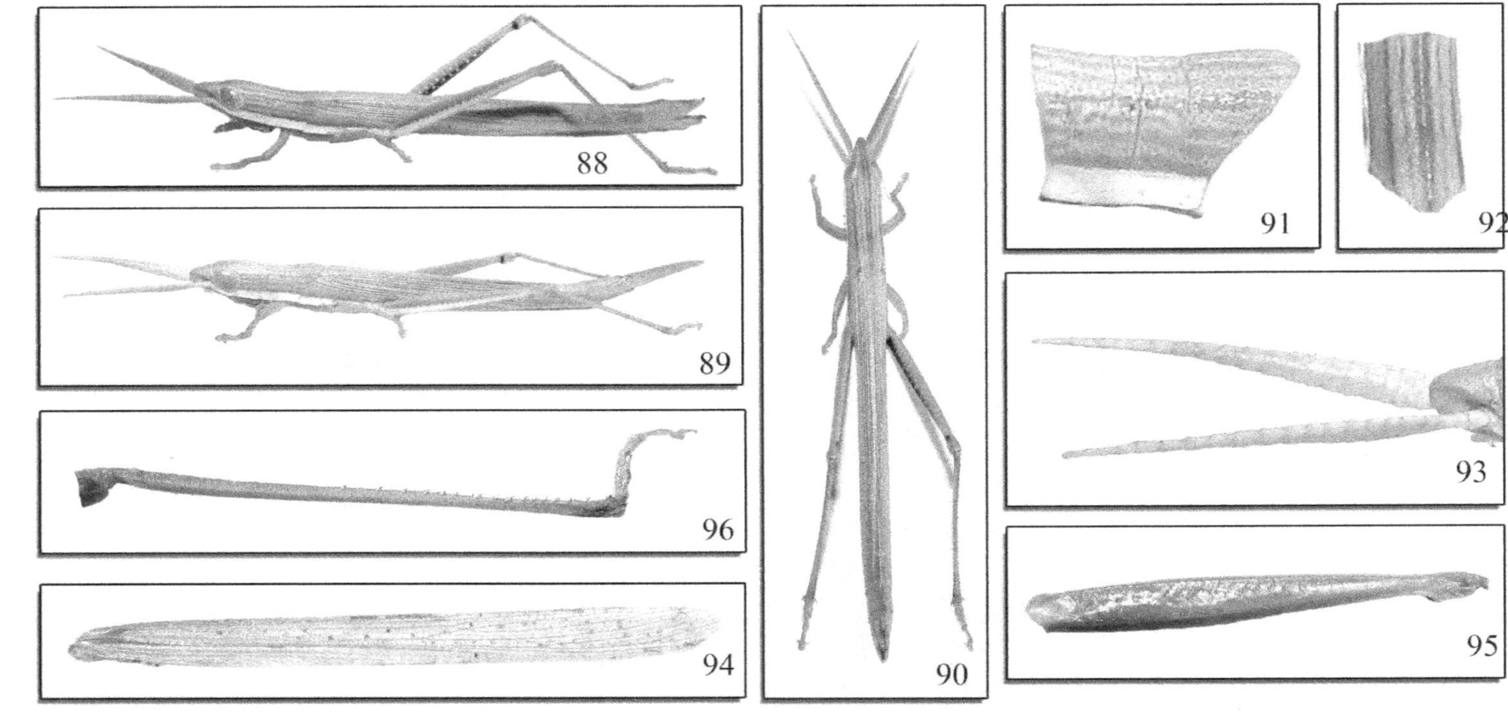

Plate 12: Figure 88: *Ischnacrida convergens* (Walk.) (Female); Figure 89: *I. convergence* (male); Figure 90: *I. convergence* (Dorsal view); Figure 91: Lateral view of Pronotum; Figure 92: Dorsal view of pronotum; Figure 93: Antennae; Figure 94: Tegmina; Figure 95: Femur; Figure 96: Tibia

Family–Acrididae

Subfamily–Oedipodinae

***Dittopternis venusta* (Walker, 1870) (Plate 13, Figures 97 to 140)**

Body brown; eyes semi round; frontal side of face laterally compressed; Head and pronotum granulose; posterior margin of matozona rectangular with tip rounded off; intercalary vein weakly serrated triangular at tip; lateral lobes of pronotum (Figures 100 and 101) laterally compressed and rounded behind, madian carina slight upwardly directed, lateral carina incomplete; tegmina (Figure 102) brown, with grey coloured patches; legs brown, pointed spines present on lateral side of fore and middle legs tibia, hind legs femora (Figure 103) short stout, black spot present on it; knee black, pale coloured ring present at basal region of tibia (Figure 104), remaining part of tibia blue coloured; 9-10 backwardly directed spines present on lateral side of tibia. In female, antennae 16 mm long; pronotum 6 mm long; thorax 7 mm long; abdomen 14 mm long; hind femora 13 mm long; hind tibia 11 mm long; tegmina 14 mm long semiround at tip.

Family–Acrididae

Subfamily–Oedipodinae

***Dociostaurus decisus* Walk, 1871 (Plate 14, Figures 105 to 113)**

Brownish; black strip with triangular markings present on the dorso lateral side of the pronotum; tegmina (Figure 111) spotted, blackish towards the tip; wings hyaline. Hind femora (Figure 112) brownish on the outer side; tibia (Figure 113) red yellowish towards the base; near about 12 black spines present on the lateral side of the hind tibia. Head 4 mm long; antennae 12 mm long, scape 1.2 mm long and 0.9 mm wide, pedicel 0.8 mm long and 0.6 mm wide, flagella 10 mm long; pronotum 3 mm long; meso and meta thorax 4 mm long; abdomen 10 mm long; tegmina 17 mm long; wing expanse 36 mm; antennae 7 mm long. This species reported in all three districts, only one generation is possible in a year.

They feed on crops like fenugreek, tomato, green gram, wheat.

Flagellar formula: 1L/W = 2, 4L/W = 1.25, 7L/W = 4, A = 2.4

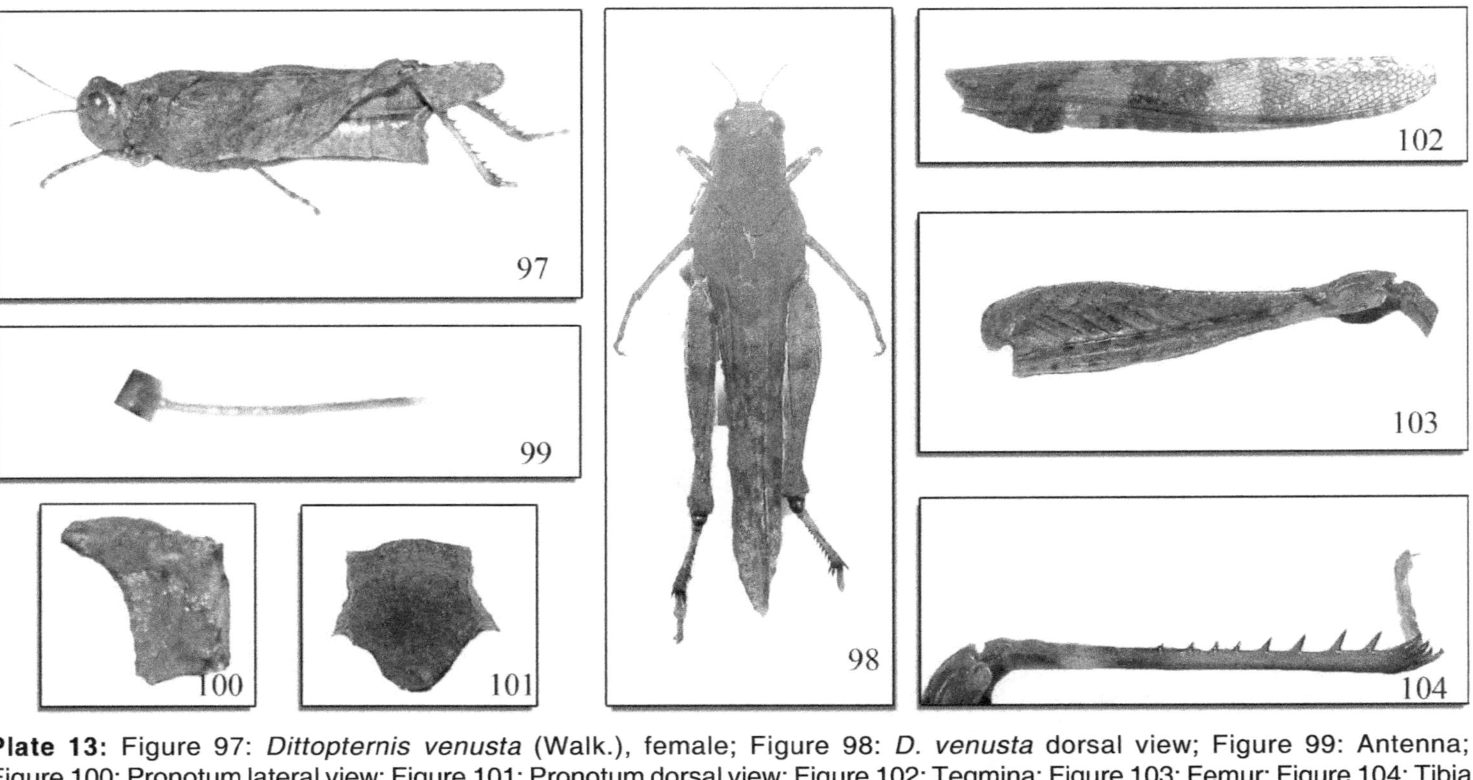

Plate 13: Figure 97: *Dittopternis venusta* (Walk.), female; Figure 98: *D. venusta* dorsal view; Figure 99: Antenna; Figure 100: Pronotum lateral view; Figure 101: Pronotum dorsal view; Figure 102: Tegmina; Figure 103: Femur; Figure 104: Tibia

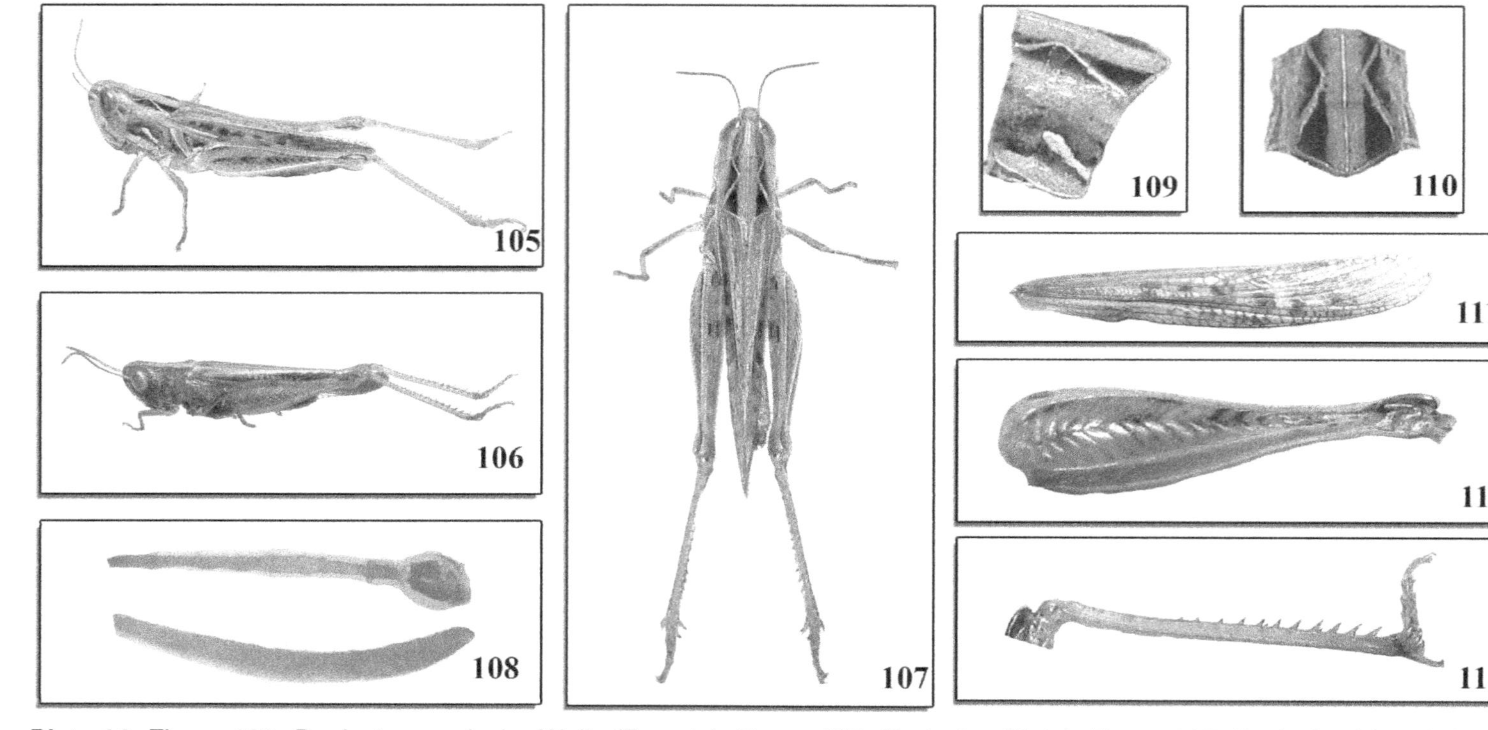

Plate 14: Figure 105: *Dociostaurus desius* Walk. (Female); Figure 106: *D. desius* (Male); Figure 107: *D. desius* (dorsal view); Figure 108: Antennae (a–Posterior side, b–Anterior side); Figure 109: Pronotum dorsal view; Figure 110: Pronotum lateral view; Figure 111: Tegmina; Figure 112: Femur; Figure 113: Tibia

Family–Acrididae

Subfamily–Oedipodinae

***Aiolopus thalassinus tamulus* (Fabricius, 1798) (Plate 15, Figures 114 to 121)**

Greenish brown also reddish tinge present on prothorax and head; antennae short pale at the base, blackish towards the tip; eyes brown, oval; a brown band present on lateral side of head just behind the eyes; fastigium of vertex with forward angle more acute; frontal ridge flat gradually narrowing; pronotum (Figures 116 and 117) short, dorsal side of pronotum reddish; tegmina long, greenish brown, white spots present on it, blackish towards the tip; hind femora (Figure 120) long, greenish, laterally flattened; hind tibia (Figure 121) yellowish towards the base, apical half red, also pale ring present on basal region; tarsus yellow; abdomen greenish, 9-10 black tipped white spine present on lateral side of tibia. In female head 5 mm long; antennae 7.7 mm, scape 0.9 mm long and 0.6 mm wide, pedicel 0.3 mm long and 0.4 mm wide, flagella 6.5 mm long; pronotum 4.5 mm long; meso and meta thorax 6 mm long; abdomen 14 mm long; tegmina 24 mm long; wing span 52 mm. In male, antennae 6 mm long; head 4 mm long; pronotum 2.5 mm long; meso and meta thorax 4 mm long; abdomen 9 mm long; tegmina 16 mm long; wing span 34 mm.

This species feed on soyabean, tur, green gram, pea, paddy, wheat.

Flagellar formula: 1L/W = 0.48, 4L/W = 0.38, 7L/W = 0.58, A = 0.48

Family–Acrididae

Subfamily–Oedipodinae

The Genus *Stauroderus* Bolivar 1879

The Genus *Stauroderus* is raised by Boliver in 1879. It shows following characters:

1. Vertex subtraingular, obtuse.
2. Foveolae superior, narrow, well marked.
3. Antennae filiform.
4. Front hardly oblique, frontal ridge carinated.
5. Pronotum with lateral carnae sharply angulated, inward before the middle.

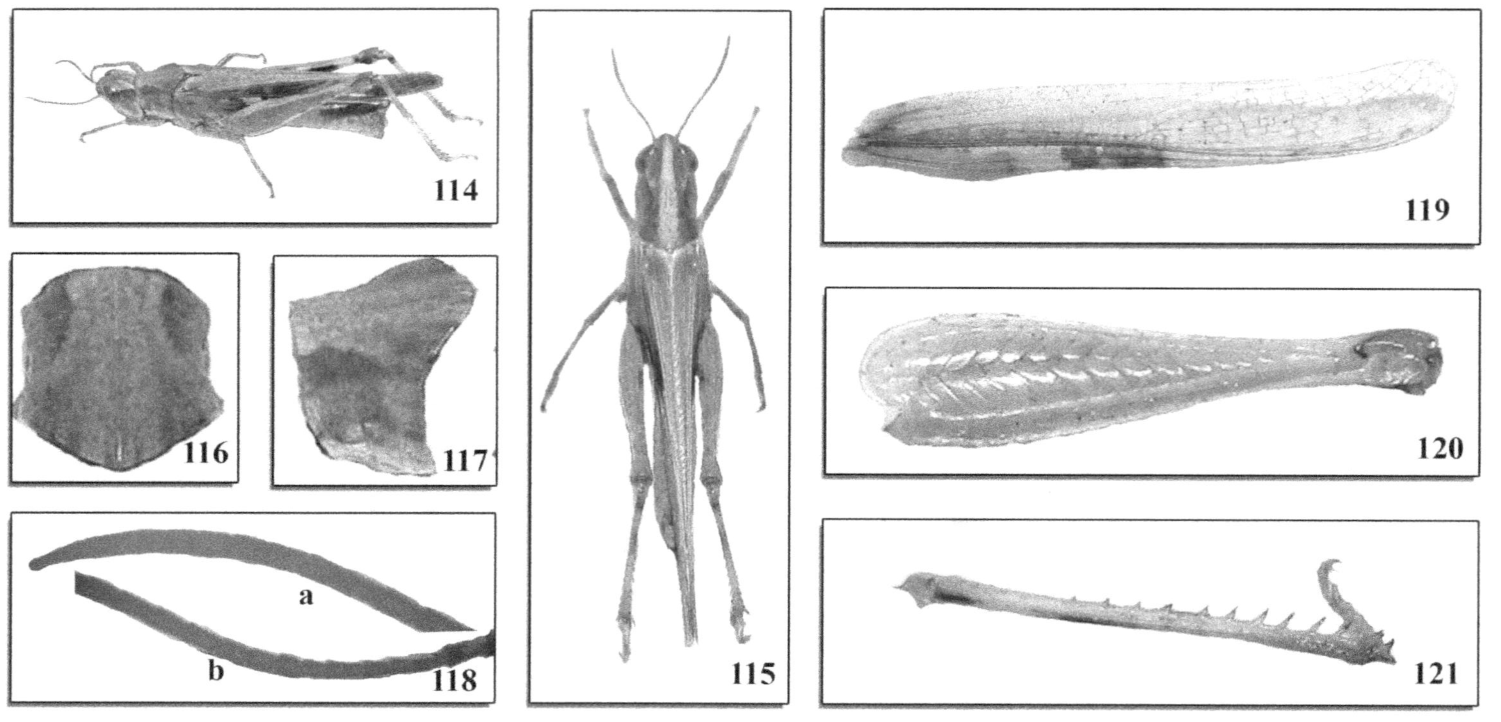

Plate 15: Figure 114: *Aiolopus thalassinus tamulus* (Fabr.) (Female); Figure 115: A. *thalssinus tamulus* (Dorsal view); Figure 116: Pronotum dorsal view; Figure 117: Pronotum lateral view; Figure 118: Antenna, (a–Anterior region, b–Posterior region); Figure 119: Tegmina; Figure 120: Femur; Figure 121: Tibia

6. Tegmina longer than the abdomen.
7. The mediastinal area slightly expanded on the costa near the base and not extending beyond the middle of the tegmina.
8. Wings hyaline.

From India only one species has been reported under this genus.

STAURODERUS INDICA SP NOV (Plate 16, Figures 122 to 130)

Female – Long, Reddish Colour

Head (Figure 127)

7 mm long (including mandible), 3mm broad, slightly pulled backward direction at ventral; eyes grey, semi oval; frontal side reddish brown, carinated; foveolar region triangular; mandible strong; rectangular foveolar cavity present at dorsolateral side of head, below this cavity present a lateral oceli; dorsal side of head present a yellowish band runs from tip of head to posterior tip of pronotum.

Antennae (Figures 125, a and b)

Longer than the head, 11.9 mm long, brown coloured; scape 1.2 mm long and 1.3 mm wide, pedicel 0.7 mm long and 0.6 mm wide, flagellum 10 mm long.

Flagellar formula 1L/W = 2.2, 4L/w=2.2, 7L/W=2, A = 2.13.

Thorax

Reddish, 7 mm long, pronotum (Figure 126) reddish with yellow band present on middorsal region, median carina white, lateral carina slightly curved inside, three sulci present on pronotum. I st sulci short, II nd sulci present only on the lateral side, III rd sulci thick present on dorsolateral region of pronotum, basal tip black, curved forward direction, it present only up to midlateral region of pronotum but II nd sulci present lateral side from dorsal yellow band to base of pronotum, meso and metathorax 6 mm long, a black and white spot present on lateral side of metathorax, just above the hind coxa.

Fore Wing (Tegmina) (Figure 128)

17 mm long, 4.5 mm wide, narrow towards the tip, broad at the costal area, longer than the abdomen; tegmina yellowish at dorsal side, reddish at basal region and greyish at tip.

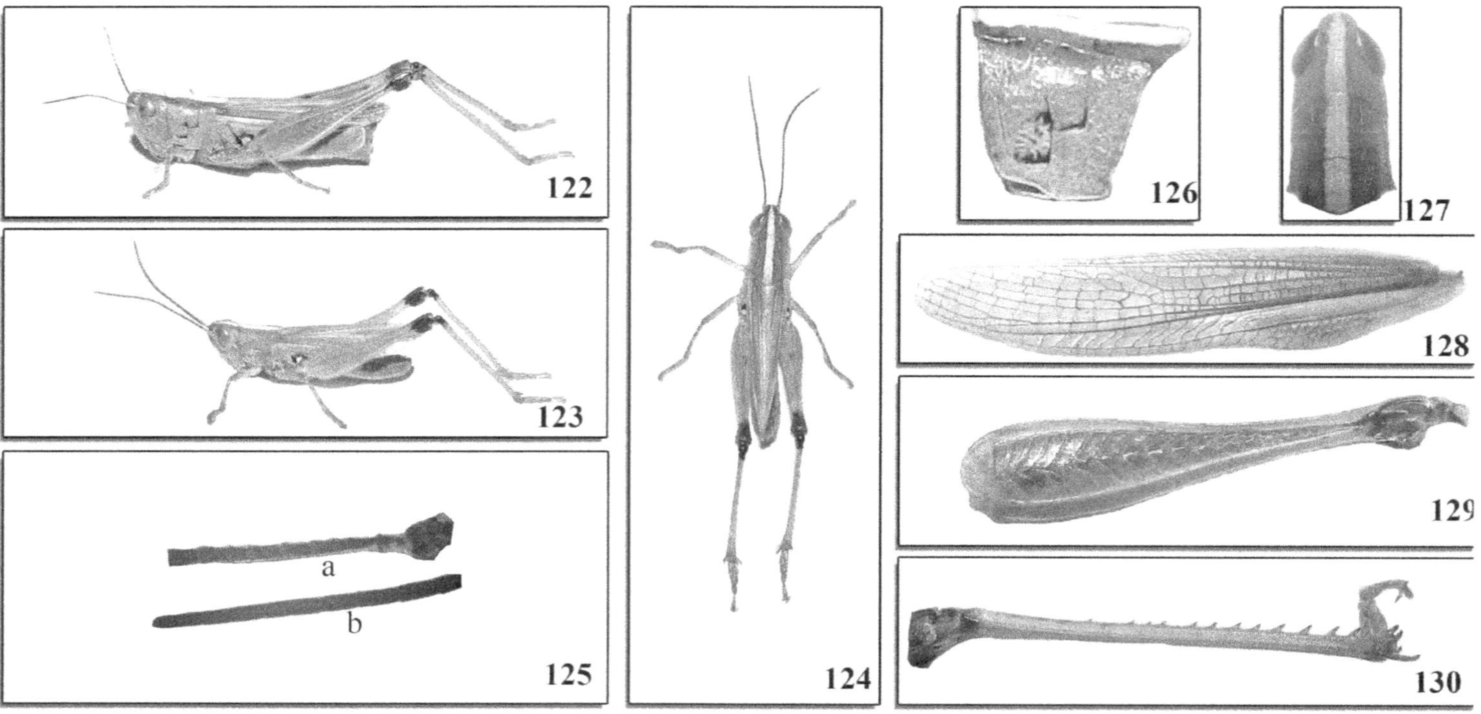

Plate 16: Figure 122: *Stauroderus indica* sp.nov. (Female); Figure 123: *S. indica* (male) Figure 124: *S. indica* dorsal view; Figure 125: Antenna (a–Posterior region, b–Anterior region); Figure 126: Pronotum lateral view; Figure 127: Head and Pronotum dorsal view; Figure 128: Tegmina; Figure 129: Femur; Figure 130: Tibia

Hind Wing

Broad, membranous, transparent greyish at tip, 16 mm long.

Fore Leg

Reddish, 10.5 mm long; coxa 1 mm long; femur reddish laterally compressed, narrow towards the base, 3.5 mm long; tibia 4 mm long, slightly broad towards the base, a row of small forwardly directed spine present at anterior region of tibia; tarsus 3 mm long, claws and arolium present at tip.

Hind Leg

34.5 mm reddish; coax 1.5 mm long; femur (Figure 129) 16 mm long, reddish colour, lateral carina of femur white, knee black; tibia (Figure 130) 12 mm long, reddish at the base and paler towards the tip; 12 black tipped backwardly directed spines presented on lateral side of tibia; tarsus 5 mm long, three segmented, IInd segment short as compare to I^{st} and IIIrd segments; at tip of tarsus present arolium and claws; dusky spot present on mid dorsal region of femur, velvety hair like bristle present on hind tibia.

Abdomen

17 mm long, reddish colour, ventral side pale and broad at base, I^{st} abdominal segment submerged in metathorax; broad, oval, auditory organ present on lateral side of first abdominal segment, I^{st} tergite 2 mm long, IInd tergite 1.5 mm long, IIIrd tergite 1.3 mm long, IVth tergite 1.2 mm long, V^{th} 1mm long, VIth tergite 0.9 mm long, VIIth tergite 0.9 mm long VIIIth tergite 1.8 mm long, IXth and X^{th} abdominal segment modified as a ovipositor, 2.3 mm long pointed and black at the tip, lower lamellae 1.2 mm long at tip, black pointed and short.

Male (Figure 123)

17.6 mm long, reddish, smaller than the female.

Head

4 mm long (including mandible), 2 mm wide; eyes oval, grey coloured; foveolar region round, oval cavity present at side of foveollae, below this cavity present a lateral ocelli, frontal ocelli yellowish; a yellowish narrow band runs over the tip of head to basal tip of pronotum.

Antennae

Antennae much longer than the head and pronotum together, 11mm long, reddish brown colour, scape 1.4 mm long, 0.7 mm wide, pedicel 0.8 mm long, and 0.7 mm wide, flagellum 8.8 mm long.

Flagellar formula- 1L/W=1.42, 4L/W=1.83, 7L/W=1.33, A=1.52.

Thorax

Reddish, 5 mm long; pronotum 3.4 mm long and 1.6 mm wide, reddish coloured; median carina white, lateral carina curved inward direction at middle region, as like a female three sulci present but basal tip of III rd sulci not black, posterior tip of pronotum convex; meso and meta thorax 4 mm long and 2.6 mm broad, mesosternal lobes yellowish broad.

Fore Wing (Tegmina) (Figure 128)

Longer than the abdomen, 11mm long, tip narrow, broader at middle region, short white band present at posterior ulnar region, yellowish at mediastinal area.

Hind Wing

Membranous, transparent, hyaline, greyish at tip, 10 mm long.

Fore Leg

Laterally compressed, 6.8 mm long, yellowish or reddish colour; coxa 0.6 mm long; femur broad towards the tip, 2.8 mm long; tibia flat and broad towards the tip, 2 mm long; tarsae 1.4 mm long claws and small arolium present at tip.

Mid Leg

7.4 mm long, reddish colour; coxa 0.8 mm long; femur laterally compressed, yellowish, 3 mm long; tibia pale colour, 3 mm long, short forward directed row of black spine presented at anterior half of tibia; tarsus 1.6 mm long, claws and arolium present at tip of tarsae.

Hind Leg

Yellowish, 15.3 mm long; coxa 1.2 mm long, pale colour; femur broad 10 mm long, knee black; tibia reddish, 0.9 mm long, pale at base, a row of pointed, black tipped spine present at lateral side of tibia; tarsus also reddish, 3.2 mm long, Ist and IIIrd segment long as compare to IInd segment, claws and arolium present at tip, velvety hair present on hind tibia.

Abdomen

8.6 mm long, yellowish red, narrow towards the tip, 9 segmented, first abdominal segment short, submerged in metathorax, a round tympanal cavity present at lateral side of first abdominal segment, I^{st} tergite 1.4 mm long, IInd tergite 1.2 mm long, IIIrd tergite 1mm long, IVth tergite 1 mm long, V^{th} tergite 1mm long, VIth tergite 0.8 mm long, VII th tergite 0.6mm long,VIIIth tergite 0.4 mm long, shown only lateral side, IXth segment modified as genetalia; cerci pointed 1mm long.

Holotype

Female India, Maharashtra, Kolhapur. Coll. Bhusnar A. R. 25 X. 2011, Head, antenna, legs, wings and abdomen mounted on the card sheet, rest body parts pinned and labelled as above. Male, India, Maharashtra, Kolhapur. Coll. Bhusnar A.R. 25. X. 2011, head, antenna, legs, wings and abdomen mounted on the card sheet, rest body parts pinned and labelled as above.

Paratype: 3♂, 3♀ sex ratio M: F 1:1, Coll. Bhusnar A.R.

Etymology

The species described in the text is reported from India for the first time hence the name *Stauroderus indica.*

Remarks

Stauroderus indica sp nov runs close to *Stauroderus bicolor* Charp by having following characters.

1. Median carina on the head and pronotum.
2. Antenna equal length in both sexes.
3. Tegmina subhyaline.
4. Legs reddish.
5. 12 small black tipped spines present on lateral side of hind tibia.

However, it differs from above species by having following features.

1. Median carina on the head pronotum which are lined with white.
2. Angulated lateral carina reddish
3. Tegmina yellowish red at base.
4. Pale and dusky spot absent on tegmina.

5. On adult female present a white sculptured spot on ventrolateral side.
6. White and black spot present on ventrolateral side of meathorax, just above hind coxa.
7. IIIrd sulci thick and curved forwardly at tip.
8. Hind knee black.
9. White short strip present at posterior ulnar area on tegmina.
10. Flagellar formula:

    Female- 1L/W= 2.2, 4L/W= 2.2, 7L/W =2, A= 2.13.

    Male–1L/W=1.42, 4L/W=1.83, 7L/W=1.33, A=1.52.

Family–Acrididae

Subfamily–Oedipodinae

***Heteropternis respondens* (Walker, 1859) (Plate 17, Figures 131-139)**

Body thick brown; dorsal part of female pronotum (Figures 135 and 136) form velvety colour at the time of breeding. Whitish or greenish band present on lateral side of pronotum. Antennae brown towards the tip and paler at the base; tegmina (Figure 137) brownish with white spot, broad towards the tip; wings slightly smaller than the tegmina and hyaline. Hind femora (Figure 138) yellowish brown irregular black spot present on the ventrolateral side of hind femora; internal spur of posterior tibia greatly unequal, lower one much longer than the other, abruptly hooked at apex, very acute; hind tibia (Figure 139) red with 10 black spines. In female, antennae 12.4 mm long, scape 1.8 mm long and 1.4 mm wide, pedicel 0.6 mm long and 0.7 mm wide, flagella 10 mm long; head 6 mm long; prothorax 3 mm long, meso and meta thorax 5 mm long; abdomen 19 mm long; tegmina 27 mm long; wing span 55 mm long; body length 30 mm. In male, antennae 10 mm; head 4 mm; prothorax 2.5 mm, meso and meta thorax 3 mm; abdomen 10 mm long.

Found association with different grasses, paddy, wheat, tur, pea, tomato.

Flagellar formula: 1L/W = 3.33, 4L/W = 2.5, 7L/W = 2, A = 2.61

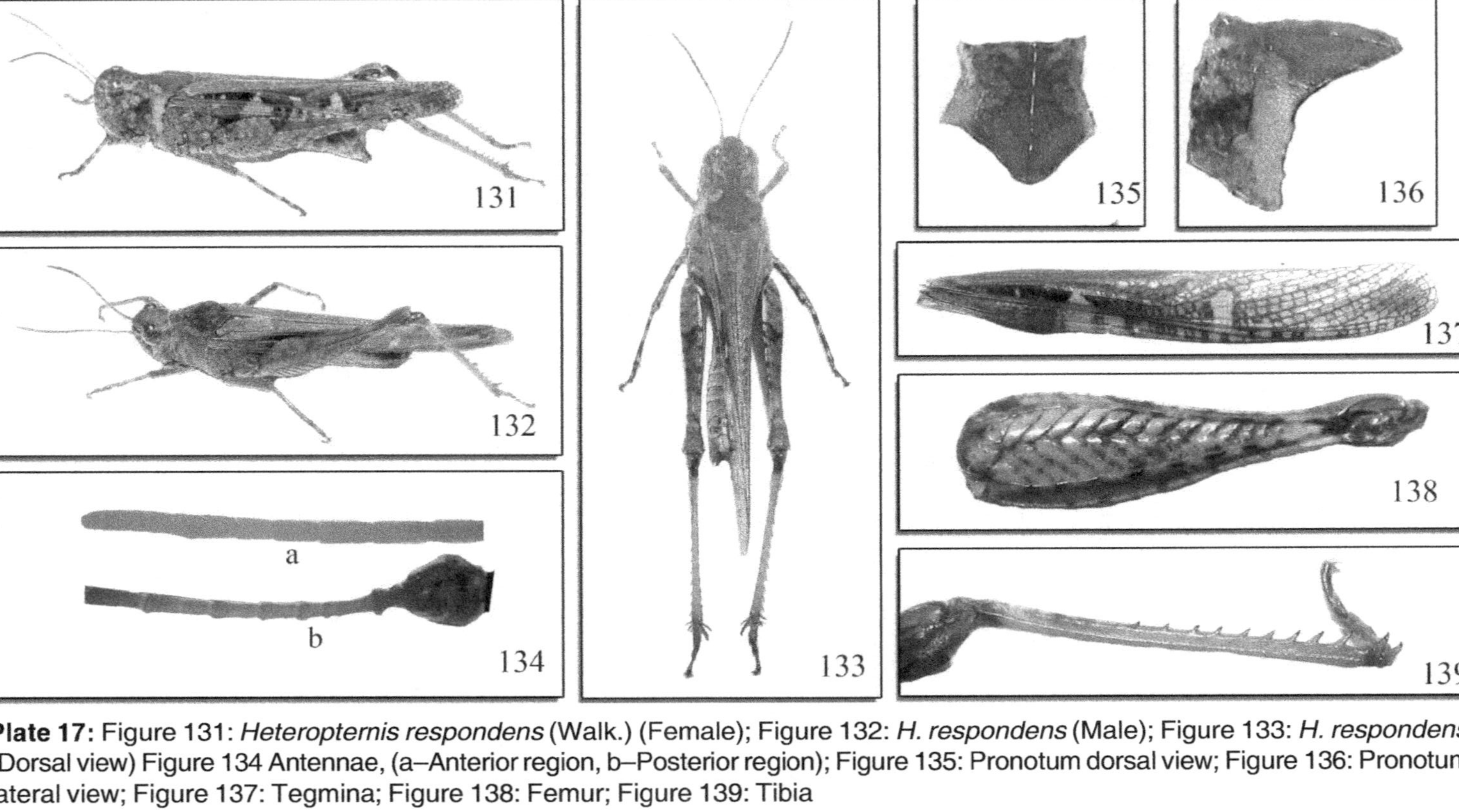

Plate 17: Figure 131: *Heteropternis respondens* (Walk.) (Female); Figure 132: *H. respondens* (Male); Figure 133: *H. respondens* (Dorsal view) Figure 134 Antennae, (a–Anterior region, b–Posterior region); Figure 135: Pronotum dorsal view; Figure 136: Pronotum lateral view; Figure 137: Tegmina; Figure 138: Femur; Figure 139: Tibia

Family–Acrididae

Subfamily–Oedipodinae

***Gastrimargus africanus africanus* Saussure, 1888 (Plate 18, Figures 140 to 148)**

Greenish, large sized body; antennae blackish at the tip and yellowish towards the base; head green; three ocelli shown with naked eyes; eyes oval, fastigium of vertex concave with median carina continuating over vertex, pronotum (Figure 144,145) moderately tectiform, brown with green, black and white patch present on lateral side of pronotum, median carina blackish at the tip. Tegmina (Figure 146) hyaline, green upto mid dorsal region, lateral side of tegmina blackish brown with broad white spots; semitransparent at the tip. Hind femora (Figure 147) green, blackish faint marks present on the base of tegmina at the dorsal side; tibia (Figure 148) red, yellow band present at the base, 10 to 12 black tipped spines present on the lateral side of hind tibia. In female, antennae 15.6 mm, scape 1.7 mm long and 1.5 mm wide, pedicel 0.9 mm long and 0.8 mm wide, flagella 13 mm long; head 8 mm; prothorax 4 mm; pronotum 9 mm; meso and meta thorax 6 mm; abdomen 20 mm long; wing span 73 mm. In male, antennae 10 mm; head 6 mm; prothorax 3 mm, meso and meta thorax 4 mm; abdomen 12 mm; tegmina 24 mm; wing span 50 mm; body length 22 mm long. This species found throughout the year. It is a pest of wheat, maize, green gram, tur, paddy, etc. and found association with long grasses.

Flagellar formula: 1L/W = 3, 4L/W = 1.83, 7L/W = 2.16, A = 2.33

Family–Acrididae

Subfamily–Oedipodinae

The Genus–*Pternoscirta* De Hann 1842

The genus *Pternoscirta* is raised by De Hann in 1842. It shows following characters:

1. Head short, somewhat rugose, the carina bounding the frontal ridge not very prominent, strongly approximating between the antennae.
2. Tegmina brown, varied with paler marking on the basal half, then with a large subhyaline patch, and more or less varied with

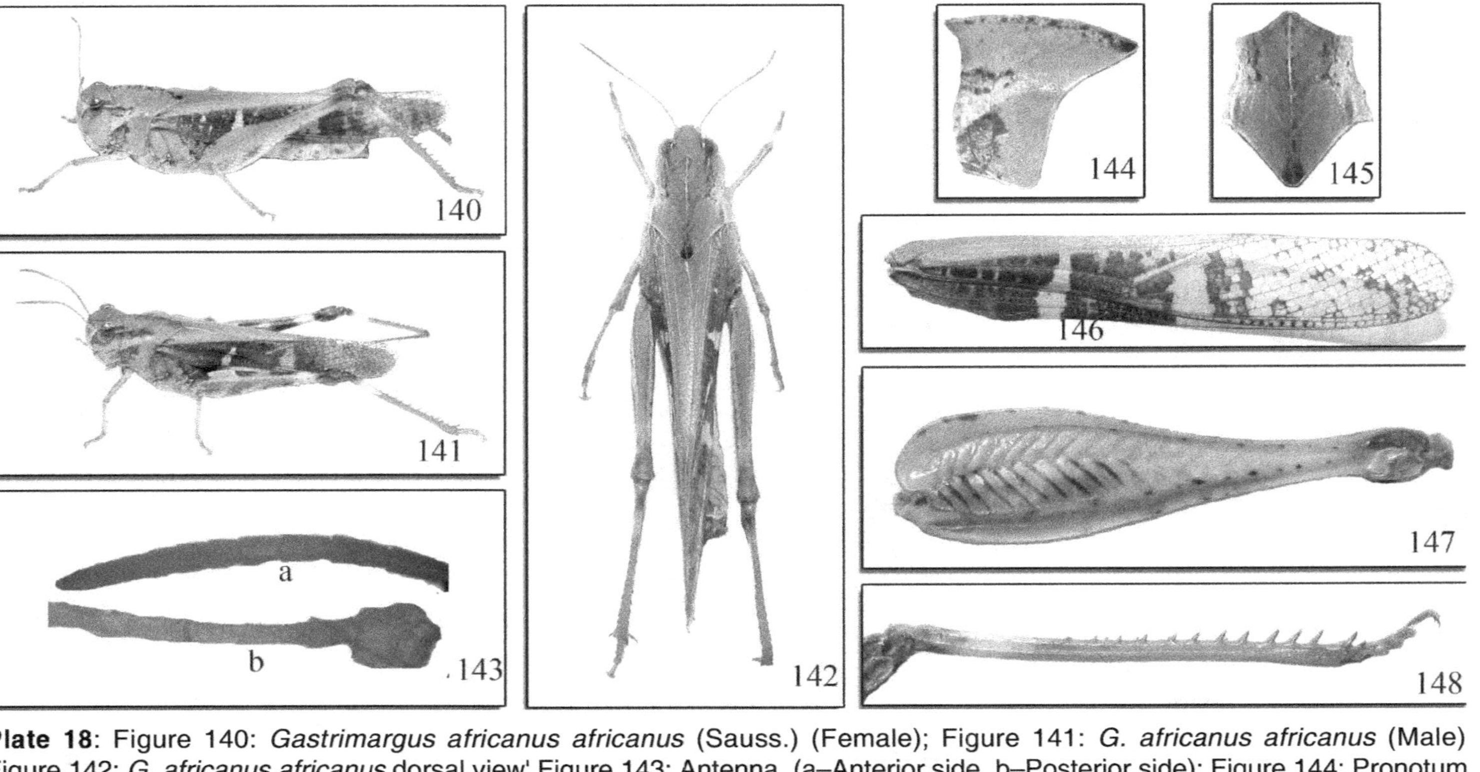

Plate 18: Figure 140: *Gastrimargus africanus africanus* (Sauss.) (Female); Figure 141: *G. africanus africanus* (Male) Figure 142: *G. africanus africanus* dorsal view' Figure 143: Antenna, (a–Anterior side, b–Posterior side); Figure 144: Pronotum lateral view; Figure 145: Pronotum dorsal view; Figure 146: Tegmina Figure 147: Femur

brown and subhyaline beyond, the cells being arranged in oblique quadrialaterals, angulated towards each other at the longitudinal nervures.

3. Wings often coloured at the base, with no central blak band, hyaline beyond the middle, with the tip dusky.
4. Legs and under surface pilose.

From India, two species have been reported under this genus.

PTERNOSCIRTA SAHYADRICUS SP NOV (Plate 19, Figures 149 to159)

Female (Figure 149)

Head

4.7 mm long and 2 mm wide, yellowish; eyes oval grey; interocellar distance 0.8 mm, frontocellar space 1.6 mm; two ocelli present on latteral side of tempora; frontal ridge narrow towards the base and broad towards the tip, mandible toothed, mouth parts yellowish.

Antenna (Figure 156)

Antenna longer than the head, thick and brown towards the tip. Scape 1 mm wide, 0.6 mm long, pedicel 0.6 mm long, 0.6 mm wide, flagellum 8 mm long, Total length 9.6 mm long.

Flagellar formula: 1L/W=1.66, 4L/W = 3.5, 7L/W = 1.6 A= 2.25

Thorax

Yellowish unicolour, 7 mm long and 4 mm wide; Pronotum (Figure 52, 53) triangular towards the tip, median carina well marked and rised upward, lateral carina not shown clearly; three sulci present on pronotum, first sulci present only on lateral side, 2nd and 3rd sulci complete, 2nd sulci longer than the 3rd sulci.

Fore Wing (Tegmina) (Figure 153)

Yellowish towards the base and greyish narrow towards the tip; tip triangular, 3 mm broad and 19 mm long.

Hind Wing

Membranus, 11 mm broad and 19 mm long, yellowish towards the tip and reddish towards the base, 19 mm long and 3 mm wide.

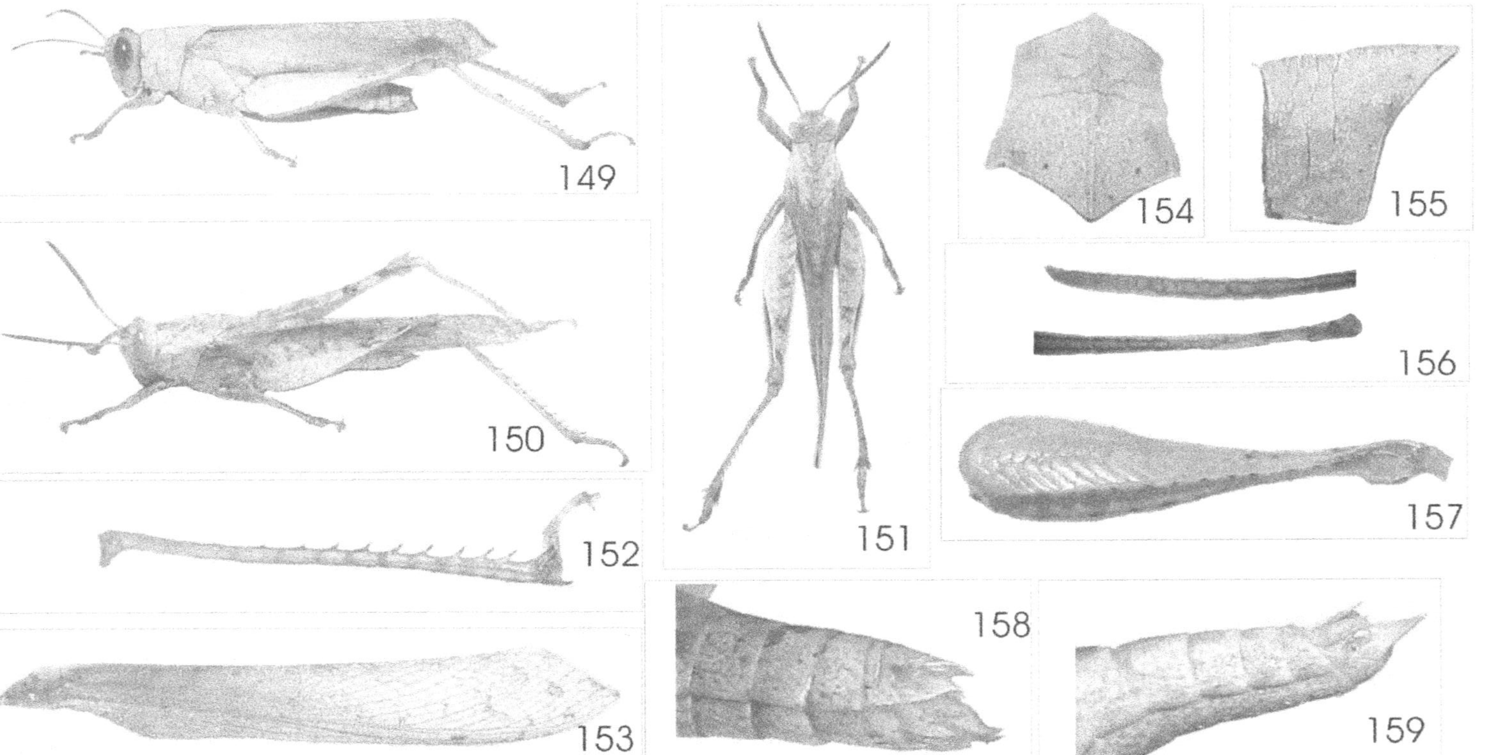

Plate 19: Figure 149: *Pternochirta sahyadricus* sp. nov. (Female); Figure 150: *P. sahyadricus* sp. (Male); Figure 151: *R. sahyadricus* dorsal view; Figure 145: Pronotum dorsal view; Figure 155: Pronotum lateral view; Figure 156: Antenna (a–Anterior side, b–Posterior side); Figure 157: Femur; Figure 152: Tibia; Figure 153: Tegmina; Figure 158: Female genitalia; Figure 159: Male genitalia

Fore Leg

Yellowish grey colour; coxa 0.6 mm long, yellow; femur laterally compressed, greyish, small velveti hair like bristle present on it; tibia 2 mm long, a black tipped spine present at lateral side; tarsus brownish, at the tip present two claws and arolium.

Mid Leg

Yellowish; Coxa 0.5 mm long; femur laterally compressed, broad towards the tip, white velvety hair like bristle present on it, 3 mm long; tibia 3.5 mm long, laterally compressed, a row of black tiped spine present on ventrolateral side towards the tip, tarsus 2 mm long, at the tip present a claws and arolium.

Hind Leg

Yellowish; Coxa 1.4 mm long; femur (Figure 157) 12 mm long, yellow, ventrolateral carina strong, ventral side flat, brown patches present on the dorsal side of femora, they shown faint towards the base; tibia (Figure 152) thick, 10 mm long, reddish or greenish, a row of 10 black tipped spines present on lateral side of tibia. Calcaria black and slightly curved, tarsus reddish, 3 mm long, claws and arolium present at tip.

Abdomen

Abdomen 10 segmented, 11 mm long, yellowish colour I^{st} segment submerged in metathorax, tympanal cavity around. Present at lateral side of first abdominal segment. I^{st} tergite 1.8 mm long, IInd tergite 1.6 mm long, IIIrd tergite 1.6 mm long, VIth tergite 1.4 mm; V^{th} tergite 1.3 mm long. VIth tergite 1.1 mm long, VIIth 1.0 mm long, VIIIth 1.2 mm long; ovipositor (Figure 158) short, upper lamillae slightly curved downword direction, bifurcated, pointed, tip black, short, 1.6 mm long, cerci short, 0.4 mm long yellow coloured.

Male (Figure 150)

Head

Yellowish coloured, 3 mm long and 2 mm broad; frontal ocelli clearly shown, eyes oval, round.

Antennae

Broad and thick brown towards the tip, yellowish towards the base, longer than the head and pronotum together. Scape 0.6 mm long, 0.7 mm wide; pedicel 0.5 mm wide, 0.6 mm long and flagellum 6 mm long.

Flagellar formula.–1L/W=1.5, 4L/W= 1.25, 7L/W=1.2 A = 1.23.

Thorax

Yellowish, sometime pinkish; pronotum slightly concave at frontodorsal side and triangular at postero-dorsal side, median carina slightly directed upword, as like female three sulci present, pronotam 3.2 mm long and 2 mm broad, meso and metathorax 4 mm long and 2.5 mm broad.

Fore Wing (Tegmina)

Forewing 17 mm long, thick yellowish towards the base and greyish towards the tip pointed, triangular, 3mm broad.

Hind Wing

Membranus, 17 mm long, yellowish towards the tip and pinkish - reddish towards the base.

Fore Leg

Greyish or yellowish, 7.3 mm long; Coxa short 0.3 mm long, femur broad 3mm long, narrow towards the base; tibia 2.4 mm long, pale coloured, a row of pointed spine present on ventrolateral side of tibia; tarsus 1.6 mm long, first two segment very short 3rd segment tip present claws and arolium; white hair like bristle presented on whole legs.

Mid Leg

7.9 mm long; white hair like bristle present on whole legs; Coxa 0.6 mm long yellowish; femur 3 mm long, greyish colour; tibia 2.3 mm long, yellowish grey colour, tarsus 2 mm long, greyish, similar to the fore leg tarsus.

Hind Leg

Yellowish, 23 mm long; coxa 1mm long, yellowish; femur stout yellowish, 10 mm long, brown patch present behind the extremity, just behind the brown patch present a group of brown spot on dorsal side of femur, on ventral side of hind femur brown spots are present; tibia 9 mm long, pale colour, a row of 10 black tipped spine present on lateral side of tibia, two pairs of calcaria present at the tip of tibia; tarsus 3mm long.

Abdomen

Yellowish, 11mm long, narrow towards the tip 9 segmented, tympanal organ present on lateral side of first abdominal segment. Ist tergite 1.1 mm long, IInd tergite 1mm long, IIIrd 1.2 mm long, IVth 1.4 mm long, Vth 0.9 mm long, VIth tergite 0.9 mm long, VIIth tergite 0.6 mm long, VIIIth tergite 0.5 mm long, XIth tergite modified as a genetalia (Figure 159), supra anal lamina 1mm long, yellowish; cerci short pointed, 0.7 mm long, lower lamina yellow, 1.3 mm long, pointed.

Holotype

Female, India, Maharashtra, Kolhapur. Coll. Bhusnar A.R. 25. III. 2011, Head, antenna, legs, wings and abdomen mounted on the card sheet, rest body parts pinned and labelled as above. Male, India, Maharashtra, Kolhapur. Coll. Bhusnar A.R. 7. IV. 2011, head, antenna, leg, wing and abdomen mounted on the card sheet, rest body parts pinned and labelled as above.

Paratype

2♂, 3♀ sex ratio M: F 1: 1.5, Coll. Bhusnar A.R.

Etymology

The species described in the text is reported from Sahyadri ranges of Western Maharashtra hence the name *Pternoscirta sahyadricus.*

Remarks

Pternoscirta sahyadricus sp nov runs close to *Pternoscirta caliginosa* De Hann. by having following characters.

1. Head with scutelum of the vertex strongly constructed.
2. Pronotum with median carina strongly marked.
3. Abdomen yellowish.
4. Tegmina yellowish grey.
5. Hind femora yellow with brown band.

However, it differs from above species by having following characters.

1. Three sulci present on pronotum, first sulci present only on lateral side, second sulci complete.
2. Wings yellowish towards tip and reddish towards the base.
3. 10 black tipped spines present on each side of tibia.
4. Tibia yellow with brown band.
5. Antennae brownish towards the tip and pale at base.

6. Flagellar formula.

   Female-1L/W=1.66, 4L/W = 3.5, 7L/W = 1.6 A= 2.25

   Male- 1L/W=1.5, 4L/W= 1.25, 7L/W=1.2 A = 1.23.

Family–Acrididae

Subfamily–Oedipodinae

***Trilophidia annulata* (Thunberg, 1850) (Plate 20, Figures 160 to 168)**

Dark brown or thick grey; on dorsal side of pronotum (Figure 164, 165) two small teeth like out growth present; eyes semi oval, brown; tegmina (Figure 166) brown, broad at the base; two white bands present on the ventral side at the tip of femora (Figure 167); on tibia (Figure 168) also two white band present, 8-9 black tipped spines present on the lateral side of tibia. In female, head 4 mm; antennae 15.6 mm, scape 1.2 mm long and 0.9 mm wide, pedicel 0.7mm long and 0.6 mm wide; prothorax 2 mm long; pronotum 3 mm long; meso and meta thorax 9 mm; abdomen 9 mm long; tegmina 12 mm long; wing span 26 mm; body length 16 mm. This is a pest of paddy, pea, tomato and found associated with green grasses.

Flagellar formula: 1L/W = 1.75, 4L/W = 1.25, 7L/W = 1.66, A= 1.55

Family–Acrididae

Subfamily–Hemiacridinae

***Spathosternum prasiniferum prasiniferum* Walk. (Plate 21, Figures 169 to 176)**

S. prasiniferum prasiniferum body colour greenish brown; head greenish; pronotum (Figure 171,172) with lateral bands, lateral carina brown, slightly raised and having a distinct median carina; on lateral side of pronotum present green, whitish and brown band. Tegmina (Figure 174) greenish brown, base pale colour, black thick spot present at central area. Tip of tegmina broad; femora green coloured; tibia pale with 10-11 black spines. Male cercus small, not bifurcated at apex; male supra anal not as above Pronotum distinctly tricarinate; fastigium of vertex not more praduced before the eyes, prosternal tubercle spathulate. In female, head 4 mm long, antennae 7.8 mm long, scape 0.9 mm long and 0.8 mm wide, pedicel 0.9 mm long and 0.5 mm wide, flagella 6 mm long; prothorax 2 mm long; pronotum 3 mm long; meso and meta thorax 4 mm long;

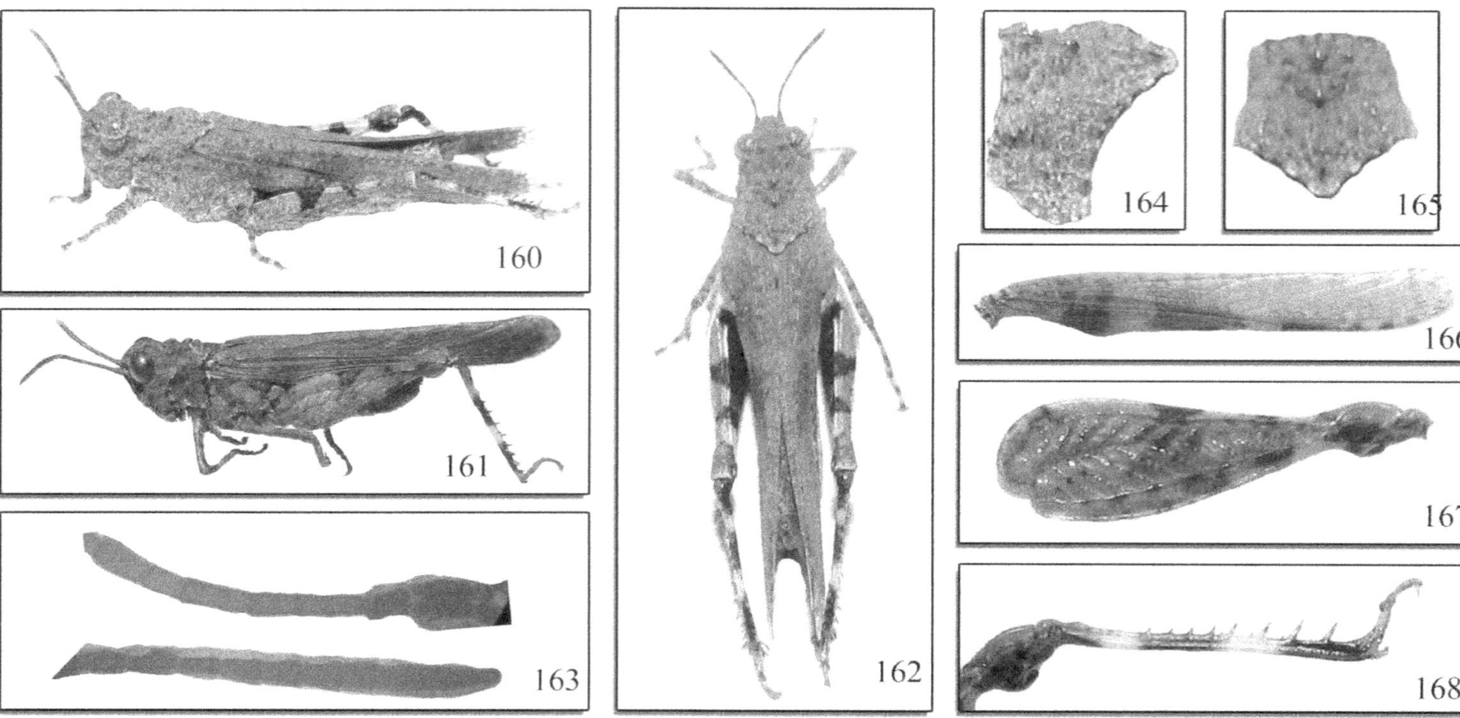

Plate 20: Figure 160: *Trilophidia annulata* (Th.) (Female); Figure 161: *T. annulata* (Male); Figure 162: *T. annulata* dorsal view; Figure 163: Antennae (a–Posterior side, b–Anterior side); Figure 164: Pronotum lateral view; Figure 165: Pronotum dorsal view; Figure 166: Tegmina; Figure 167: Femur; Figure 168: Tibia

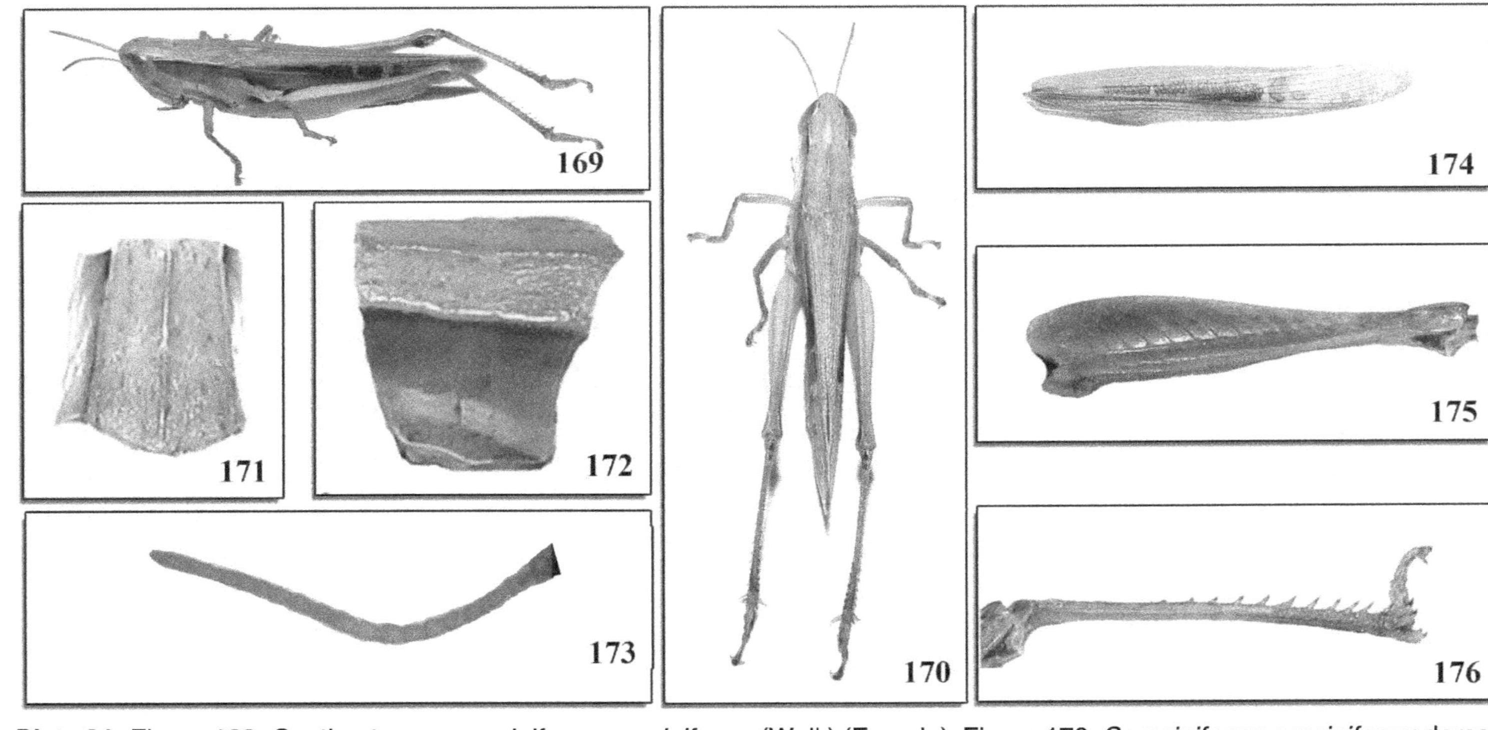

Plate 21: Figure 169: *Spathosternum prasiniferum prasiniferum* (Walk) (Female); Figure 170: *Sprasiniferum prasiniferum* dorsal View; Figure 171: Pronotum lateral view; Figure 172: Pronotum dorsal view; Figure 173: Antenna; Figure 174: Tegmina; Figure 175: Femur Figure 176: Tibia

abdomen 9 mm long; tegmina 14 mm long; wing span 30 mm and in male head 4 mm long; prothorax 2 mm long; pronotum 3 mm long; meso and meta thorax 4 mm long.

It is a pest of wheat, tomato, fenugreek and paddy.

Flagellar formula: 1L/W = 0.75, 4L/W = 0.8, 7L/W = 1.4, A= 0.98

Family–Acrididae

Subfamily–Hemiacridinae

THE GENUS *HIEROGLYPHUS* KRAUSS 1877

The genus *Hieroglyphus* is raised by Krauss in 1877. It shows following characters:

1. Size large, head very large, eyes wide apart.
2. Fastigium of the vertex short, convex, rounded in front.
3. Costal ridge broad, entire slightly sloping.
4. Antennae slender, filiform, much longer than the head and pronotum together in the male and as long or longer in the female.
5. Pronotum as broad as the head, with the sulci, very strongly marked usually with black spine.
6. Prosternal tubercle acute; meso and metasternal lobes more or less widely separated or contiguous in the male.
7. Geniculor lobes of hind femora pointed or rounded.
8. Hind tibia with 8 or 10 spines on the outer carina and 10 on the inner.
9. Tegmina subhyaline, very thickly reticulated towards the base and with the costa only slightly expanded.
10. Wings hyaline, rather long, narrow and pointed.
11. Male with the cerci long, pointed or obtuse, with a tooth on the inner side; subgenital lamina long, conical pointed.
12. Female with the genitalia valves short, thick, curved the upper ones very broad with the outer margin crenulated the lower pointed and armed with a tooth beyond the middle.

From India, seven species have been reported under this genus.

Hieroglyphus banian Fab. (Plate 22, Figures 177 to 187)

Female (Figure 177)

44 mm long, green, larger than the male.

Head

5.6 mm long and 5.5 mm wide, green; eyes reddish brown, oval in shape; three ocelli present, interocellar distance 3 mm; frontocellar space 4 mm; one ocelli present on middle region of frontal ridge and other two ocelli present lateral side on tempora, carina narrow towards the eyes; mandible toothed strong and; labrum broad, reddish brown.

Antenna (Figure 184)

Antenna much longer than the head, 20 mm long, 24 segmented, green or brownish with yellow bands, placods absent; scape brown, 1 mm long; pedicel green, 0.9 mm long; flagellum 18 mm long, 22 segmented.

Flageller Formula: 1 L/W = 1.3, 4 L/W = 1.6, T L/W = 1.8, A = 1.56

Thorax

Green, 14 mm long and 8 mm wide; pronotum (Figure 182,183) smooth, not unicoloured median carina of pronotal disc weak but entire divided by three sulci; 1st Sulcus present only laterally, 2nd medially, 3rd and posterior, 2nd and 3rd slightly curved. pronotum 8.7 mm long larger than the head; meso and metasternal lobe less separated by a grove.

Fore Wing

Fore wing thick, green dark on base and faint on tip, 31 mm long and 6 mm wide, slightly shorter than the hind wings, subhyaline; densely reticulated. Mediastinal vein 26 mm long. Scapular 34 mm long; middal radial 35 mm long; posterior ulnar 79 mm, anterior ulnar 33 mm; anal 29 mm and axillary 21 mm.

Hind Wing (Figure 185)

Green, transparent, slightly longer than the tegmina, 32 mm long and 18 mm wide, scapular vein 34 mm, middle radial vein 36 mm long, posterior radial 35 mm, anterior ulnar 33 mm, posterior ulnar 22 mm anal 34 mm and axillary 31 mm long.

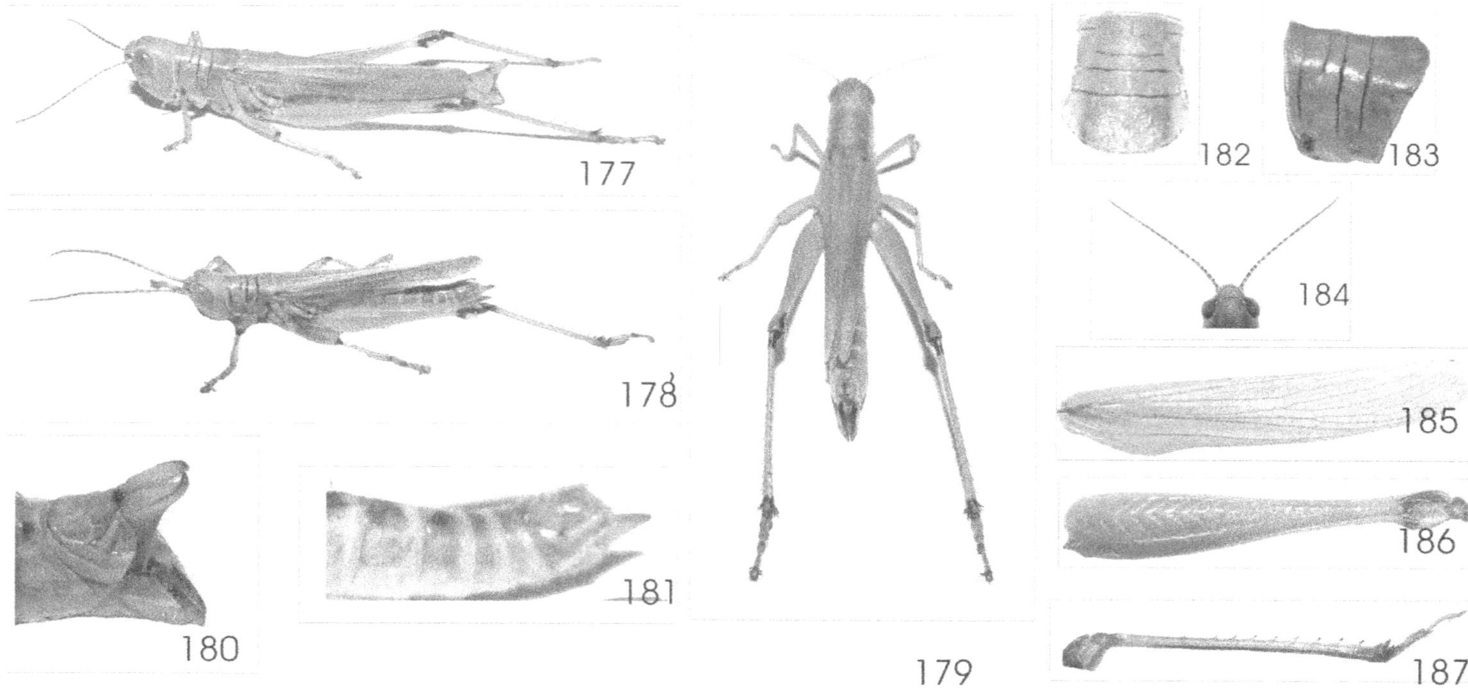

Plate 22: Figure I 77: *Heiroglyphus banian.* (Female); Figure 178: *H. banian* (Male); Figure 179: *H. Banian* dorsal view; Figure 180: Abdomenal tip of female; Figure I81: Abdomenal tip of male; Figure I82: Pronotum dorsal view; Figure 183: Pronotum lateral view; Figure 184: Antenne; Figure 185: Tegmina; Figure 186: Femur; Figure 187: Tibia

Fore Leg

20 mm long, coxa 0.8 mm long; trochanter 0.3 mm long, black; femur 7 mm, smooth; tibia 6.8 mm long, front side tibia with 8 spines and back side with 6 spines; tarsus 5 mm long, three segmented, I^{st} and IIIrd longer than the IInd, at the tip of tarsus two hook like claws are present, in between claws arolium is also present.

Mid Leg

Slightly longer than the fore legs; coxa 1 mm; trochanter black, 0.3 mm long; femur 7.7 mm long, smooth; tibia 7 mm long, on front side of tibia 5 black spines and at back side 9 black spines are present, front sides, spine bigger than the back side is spine, tarsus 5 mm long, size of tarsal segment is like a fore leg.

Hind Leg

It is 59 mm long, coxa 4 mm long trochanter black 1 mm long; femur (Figure 186) 25 mm long; tibia (Figure 187) 20 mm long, blue colour; Tarsus 9 mm long, three segmented, I^{st} and IIIrd tarsal segment longer than the IInd tarsal segment, a apical spine and calcaria is also present, at the joint on tibia and tarsus 9 black spines are present on both side of tibia.

Abdomen

30 mm long, narrow posteriorly, 10 segmented, first segment submerged in thorax; tympanal organ situated laterally on the first abdominal segment. I^{st} tergite 4 mm, IInd tergite 4.1 mm, IIIrd tergite 4 mm, IVth tergite 3.7 mm, V^{th} 3mm, VIth 2.3 mm; VIIth 2.1 mm, VIIIth 2.3 mm, IXth and X^{th} are modified as ovipositor; short ovipositor (Figure 180), 9 mm long, curved upper and lower lamilae, upperlamillae slightly curved upward, 5 mm long, lower laminal tip curved downward, they are 4 mm long, small reduced anal cerci 1 mm long.

Male (Figure 178)

Smaller than the female.

Head

4 mm long, 3 mm wide, green, shape as like a female; eyes oval, reddish brown; three ocelli present, position as per the female; other character of head more or less similar to the female, only size is small.

Antenna

23 mm long, 24 segmented, remaining all characters of antenna more or less similar to the female.

Flageller formula: 1 L/W = 1.4, 4 L/W = 1.8, T L/W = 2.2, A = 1.8.

Thorax

16 mm long and 5 mm wide, pronotum 6 mm long and 4 mm wide. Shape as like a female, all remaining characters more or less similar to the female except sexual characters.

Fore wing: 25 mm long, 4 mm wide,

Hind wing: 26 mm long.

Fore Leg

Greenish, coxa 1 mm; trochanter 0.3 mm, femur 6.0 mm long, white hairs present on femur, tibia 5.5 mm; apical spine and calcaria absent, tarsae 5 mm long.

Mid Leg

Coxa 1 mm; trochanter 0.2 mm, femur 6 mm and tibia 6.4 mm long; apical spine and calcaria absent; tarsus 5 mm, white hairs present on whole leg.

Hind Leg

Coxa 3 mm, trochanter 0.3 mm, femur 8.4 mm and tibia 15 mm long and blue; apical spines and calcaria present;; black tipped spines arranged in two rows, three segmented tarsae with claws and arolium, 5 mm long.

Abdomen

Yellowish Green colour, 19 mm long, nine segmented scleroids, small as comparative to female, 9th segment modified as a genitalia, subgenital lamina (Figure 181) pointed 2.4 mm long; anal cerci well marked 1.7 mm long.

Holotype

Female, India, Maharashtra, Kolhapur. Coll. Bhusnar A.R. 18. IX. 2010, head antenna, legs; wing and abdomen mounted on the card sheet, rest body parts pinned and labelled as above.

Male, India, Maharashtra, Kolhapur. Coll. Bhusnar A.R. 14. IX. 2010, head, antenna, leg, wing and abdomen mounted on the card sheet, rest body parts pinned and labeled as above.

Paratype

7 ♂, 5♀ sex ratio (M:F) 1:1.4 coll. Bhusnar A.R. Oct. 2010 to Nov. 2010.

Remarks

1. Pronotum smooth.
2. Tegmina subhyaline.
3. Antenna much longer than head.
4. Pronotum black narrow lines on dorsal side.
5. Antenna greenish or Brown with yellow bands.
6. Flagellar formula:

   Female: 1 L/W = 1.3, 4 L/W = 1.6, T L/W = 1.8, A = 1.56

   Male : 1 L/W = 1.4, 4 L/W = 1.8, T L/W = 2.2, A = 1.8

Family–Acrididae

Subfamily–Hemiacridinae

***Parahieroglyphus bilineatus* [(Bolivar, 1912) Plate 23, Figures 188 to 196]**

Body green; head broad, frontal space of head reddish brown; outer side of mandible green; eyes brown oval; pronotum (Figures 191, 192) green; median carina slightly shown, lateral carina brown; three sulci present on pronotum, first sulci short, incomplete, second and third complete; fore and mid leg brown; abdomen green; hind femora (Figure 195) reddish brown; a black spot present near at extrimite; tegmina (Figure 194) short narrow towards both end; tibia (Figure 196) faint blue, 8-9 black tipped spines present on lateral side of hind tibia. Male cercus very large, trilobite, male supra anal plate broader than long, with straight side narrowing to acute angular apex. In female, antennae (Figure 193) 13 mm long; head 12 mm long; pronotum 9 mm; meso and meta thorax 10 mm long; tegmina 10 mm long; wings absent; abdomen 18 mm long; hind femora 12 mm long; hind tibia 18 mm long. In male, antennae 12 mm long; head 6 mm long; pronotum 4 mm; meso and meta thorax 6mm

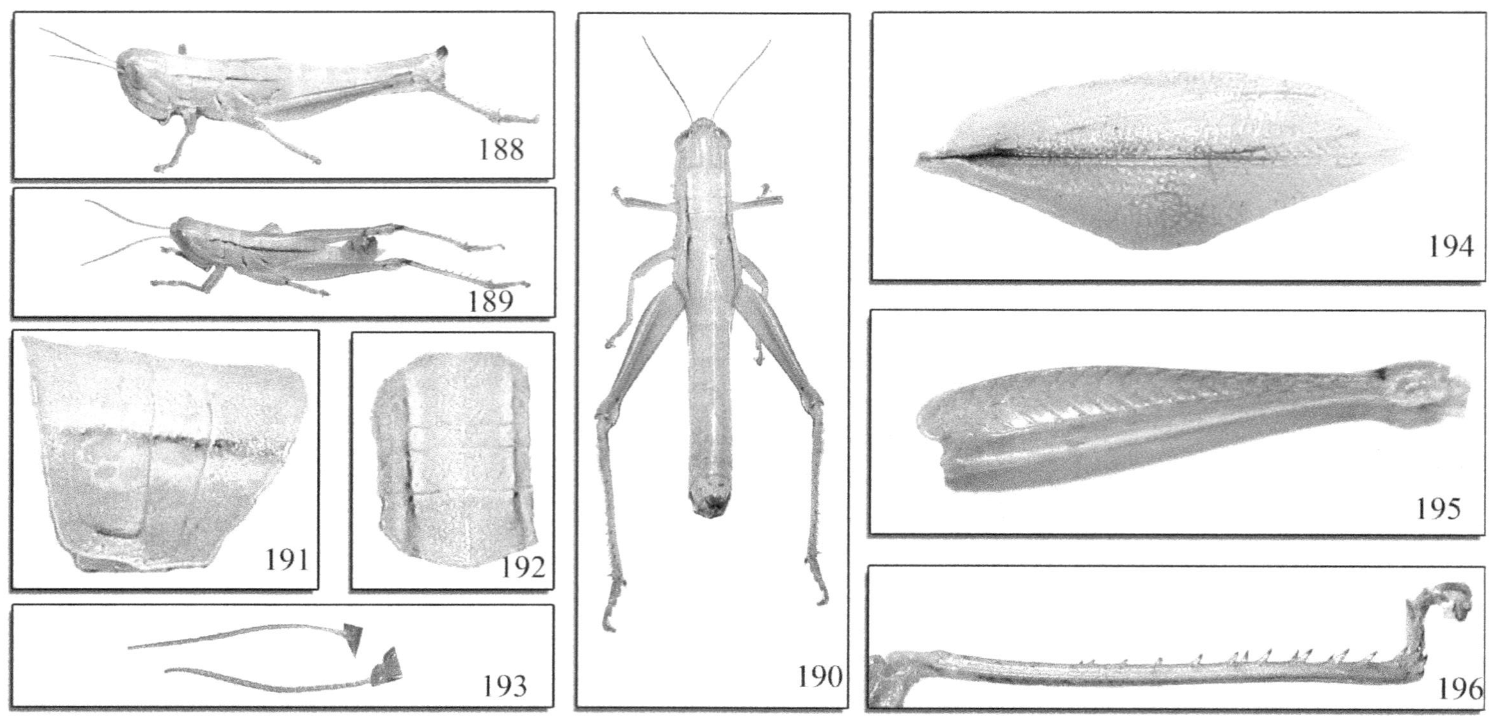

Plate 23: Figure 188: *Parahieroglyphus bilineatus* (Bol.) (Female); Figure 189: *P. bilineatus* (Male); Figure 190: *P. bilineatus* dorsal view; Figure 191: Pronotum lateral view; Figure 192: Pronotum dorsal view; Figure 193: Antennae; Figure 194: Tegmina; Figure 195: Femur; Figure 196: Tibia

long; tegmina 9 mm long; hind femora 12 mm long; hind tibia 12 mm long; abdomen 14 mm long.

Family–Acrididae

Subfamily–Oxyinae

Oxya hyla hyla **Serville, 1833 (Plate 24, Figures 197 to 206)**

Green coloured body; a brown band run over the lateral side of head, pronotum to tip of tegmina; eyes brown, oval; antennae pale coloured at the base, tip black; front side of head granulated; antennae longer than the head; pronotum (Figure 201, 202) smooth, flattened, with the median carina slightly marked or wanting, metasternal lobes continuous in both sexes, tip convex; tegmina (Figure 203) longer than the abdomen; legs yellowish green; tibia bluish, 8-9 black tipped spines present on lateral side of tibia (Figure 206); ventral surface of sub genital plate present two longitudinal ridges. Male Cercus conical or if compressed then narrowing towards apex, which is obtuse or truncate. Female subgenital plate markedly concave and toothed lateral ridges. In female, antennae 11.6 mm long, scape 1.7 mm long and 1.1 mm wide, pedicel 0.9 mm long and 0.6 mm wide, flagella 9 mm long; head 7 mm long; pronotum 6 mm long, meso and meta thorax 6 mm long; abdomen 15 mm long; tegmina 23 mm long; wing span 48 mm; hind femora 16 mm long; hind tibia 13 mm long. In male, antennae 9 mm long; head 5.6 mm long; pronotum 4 mm long; tegmina 17 mm long; wing span 36 mm; abdomen 12 mm long; hind femora 12 mm long; hind tibia 10 mm long. This species feed on paddy, wheat, brinjal, tomato, maize.

Flagellar formula: 1L/W = 1.25, 4L/W = 1.5, 7L/W = 1.4, A= 1.38

Family–Acrididae

Subfamily–Oxyinae

Oxya japonica japonica **(Thunberg, 1815) (Plate 25, Figures 207 to 215)**

Medium sized, olive green; eyes oval brown; antennae pale; a brown band run behind the antennae over the tegmina; pronotum smooth, not differentiate from head, median carina slightly shown; lateral carina semi round; dorsal side of tegmina greenish and ventrolateral side brownish, slightly pointed at the tip; hind femora (Figure 214) green, narrow; tibia

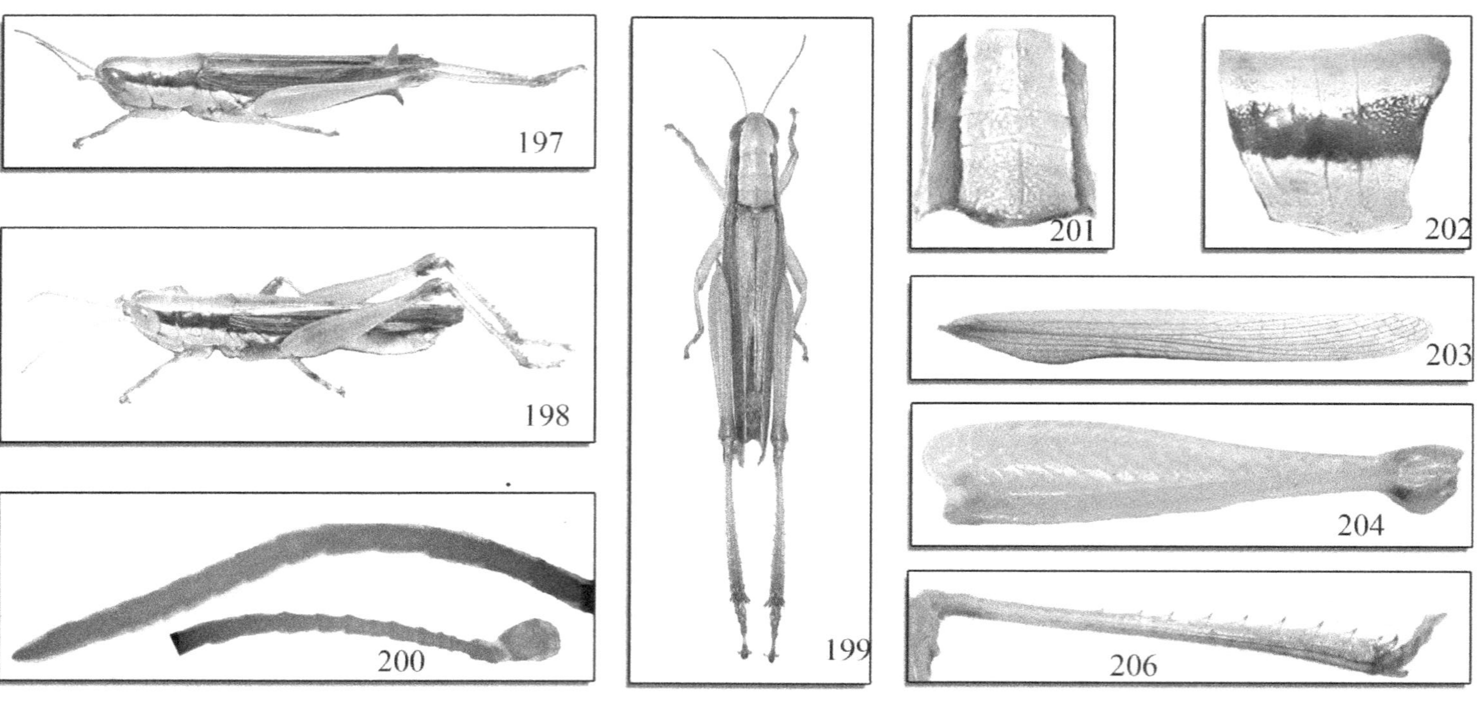

Plate 24: Figure 197: *Oxya hyla hyla* (serv.) Female; Figure 198: *O. hyla hyla* male; Figure 199: *O. hyla hyla* dorsal view; Figure 200: Antennae, (a–Anterior tip, b–Posterior tip); Figure 201: Pronotum dorsal view; Figure 202: Pronotum lateral view; Figure 203: Femur; Figure 204: Tegmina; Figure 206: Tibia

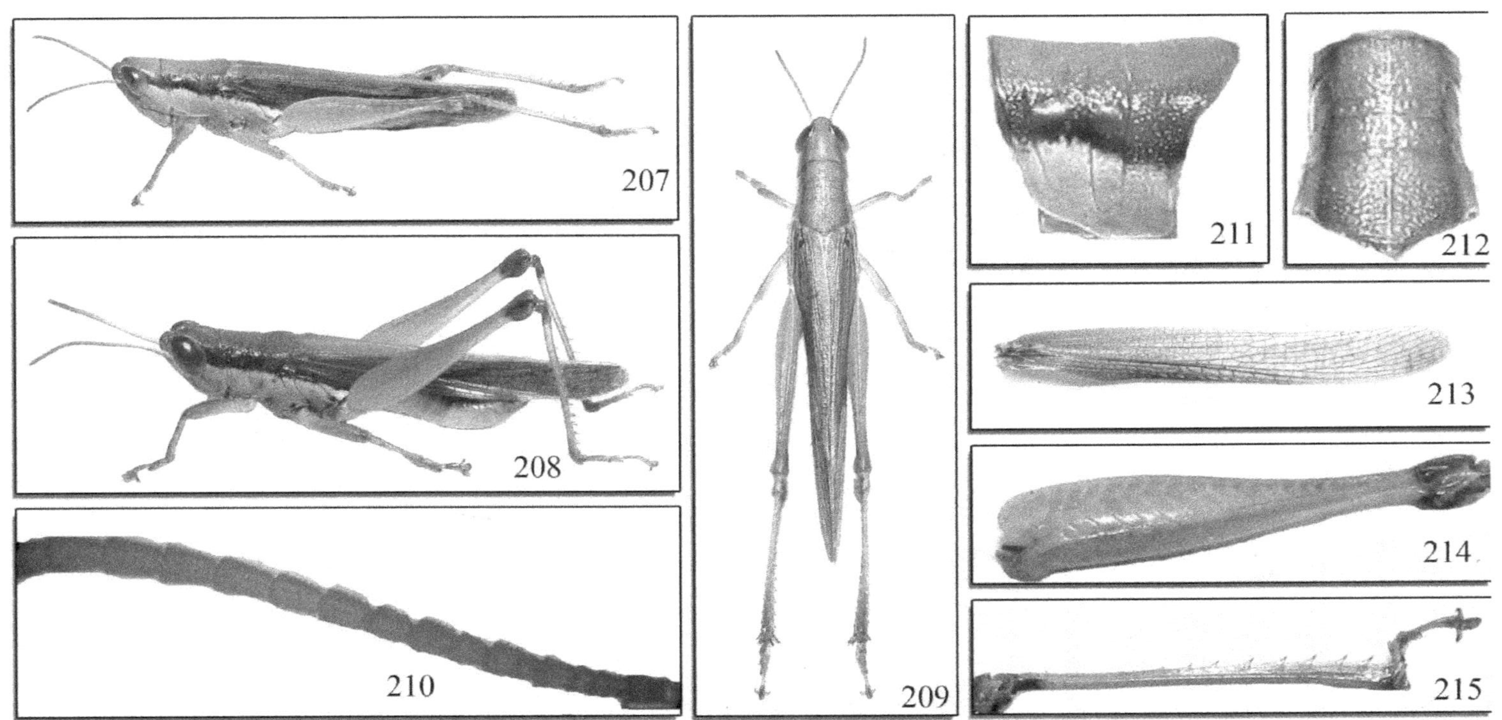

Plate 25: Figure 207: *Oxya japonica japonica* (Th.) Female; Figure 208: *O. japonica japonica* male, Figure 209: *O. japonica japonica* dorsal view; Figure 210 Antennae; Figure 211: Pronotum lateral view; Figure 212: Pronotum dorsal view; Figure 213: Tegmina; Figure 214: Femur; Figure 215: Tibia

pale, 8-9 black tipped spines present on lateral side of hind tibia (Figure 215). In female posterio ventral sclerites having one or two tooth like spines present; lateral longitudinal ridges of ovipositor with two teeth at apices. In female, head 7 mm long; prothorax 4 mm long; tegmina 23 mm long; wing span 50 mm long and in male head 6 mm long; prothorax 2.5 mm long, meso and meta thorax 5.5 mm long; abdomen 9 mm long; tegmina 16 mm long and wing span 30 mm.

It damages the crops like maize, paddy, brinjal, sugarcane.

Family–Acrididae

Sub Family–Cyrtacanthacridinae

***Patanga succincta* (Johonsson, 1763) (Plate 26, Figures 216 to 224)**

This is one of the largest species, yellowish brown; head strong with yellowish longitudinal brown bands; eyes brown, oval in shape; antennae pale; white band runs dorsally in between the eyes to central region of tegmina. Pronotum (Figure 220, 221) stout, hind margin obtus to triangular; median carina of pronotum is slightly carinated, on lateral side of pronotum present brown and white bands alternately; tegmina (Figure 222) thick, with straight venation in apical part, transverse vein forming almost right angle with principle veins, they are broad towards the tip, brown at the base, lateral margin white; wing base rosy violet; hind femora (Figure 223) strong carinated, black spots present on the dorsal side of hind femora; tibia pale, 8-10 black tipped spines present on the lateral side of tibia. In female, head 12 mm long, antennae 23 mm, scape 2.2 mm long and 2.7 mm wide, pedicel 1.8 mm long and 1.4 mm wide, flagella 19 mm long; pronotum 12 mm long; meso and meta thorax 16 mm long; abdomen 28 mm long; body length 59 mm long; tegmina 62mm long; wing span 128 mm. In male genital plate long, curved upwards, conical and apex pointed; head 10 mm long; antennae 17mm long; pronotum 6 mm long; meso and meta thorax 13 mm long; abdomen 28 mm long; body length 53 mm; tegmina 55 mm long; wing span 115 mm.

This species is associated with grasses like *Apluda mutica* L., *Heteropogon contortus* (L.) and *Themeda quadrivalvis* (L).

Flagellar formula: 1L/W = 2.12, 4L/W = 2.11, 7L/W = 1.7, A = 5.93

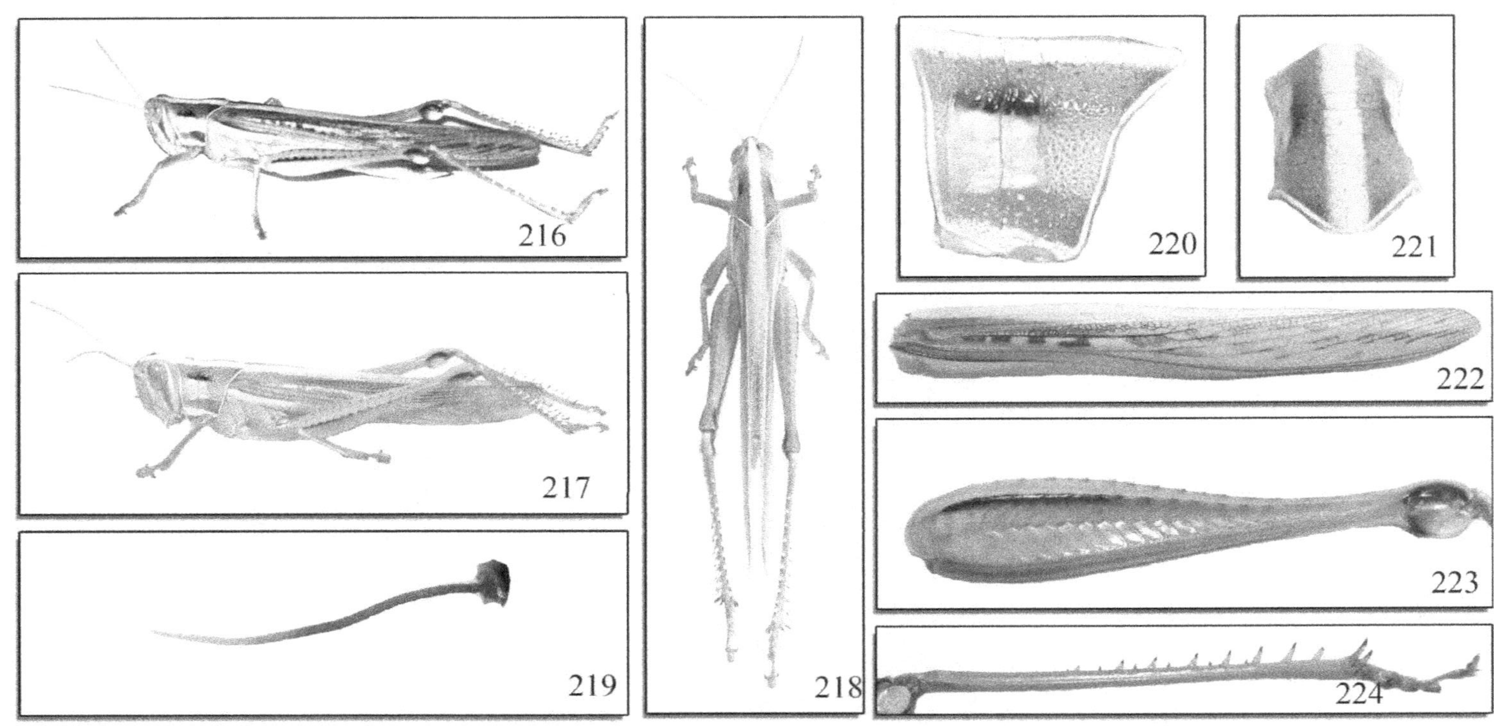

Plate 26: Figure 216: *Patanga succincta* (Johansson) Female; Figure 217: *P. succincta* (Male); Figure 218: *P. succincta* dorsal view; Figure 219: Antennae; Figure 220: Pronotum lateral view; Figure 221: Pronotum dorsal view; Figure 222: Tegmina; Figure 223: Femur; Figure 224: Tibia

Family–Acrididae

Subfamily–Cyrtacanthacridinae

***Cyrtacanthacris tatarica* (Linnaeus, 1758) (Plate 27, Figures 225 to 233)**

Antennae (Figure 228) pale; eyes oval, brown. Head broad whitish with longitudinal brown band and irregular spots. Pronotum little constricted; integument fairly rugose or dotted; prothorax is well designed, median carina is slightly carinated. A white band runs dorsaly from in between the eyes to center of tegmina; longitudinal brown band present on dorsolateral side of pronotum (Figure 229, 230), below this band is present a white band, below white band is again present brown band and below brown band there is a small narrow white band; tegmina (231) is thick, irregular, brownish spot present on it; hind femora (Figure 232) strong; tibia pale coloured, with 6-8 strong black tipped spines. In male circus with sub acute apex. In female, head 12 mm long; antennae 23.4 mm, scape 3.6 mm long and 2.3 mm wide, pedicel 1.8 mm long and 1.5 mm wide, flagella 18 mm long; prothorax 7 mm long; pronotum 14 mm long; meso and meta thorax 15 mm long; abdomen 30 mm long; tegmina 57 mm long; wing span 120 mm.

It is a pest of groundnut, soyabean, cotton, tur, jowar.

Flagellar formula: 1L/W = 1.83, 4L/W = 1, 7L/W = 1.66, A = 1.4

Family–Acrididae

Subfamily–Cyrtacanthacridinae

***Anacridium flavescence* Fabricius, 1793 (Plate 28, Figures 234 to 241)**

Reddish brown body; head large; eyes oval; antennae black; Pronotum (236, 237) with high median carina, anterior part comb shaped, lateral carina smooth; tegmina (Figure 239) long and brown with black spot; on ventrolateral carina of hind (Figure 242) femora present a row of black dots; lateral side of hind femora pale coloured; tibia (Figure 241) greenish brown, 7-9 black tipped red spines present on tibia. In female, antennae 25 mm long; head 11 mm long; prothorax 7 mm long; pronotum 13 mm long; meso and meta thorax 17 mm long; abdomen 30 mm long; tegmina 73 mm long; wing span 152 mm. *A. flavescense* feeds on jowar, paddy and different long grasses.

Flagellar formula: 1L/W = 1.32, 4L/W = 1.19, 7L/W = 1.66, A = 1.39

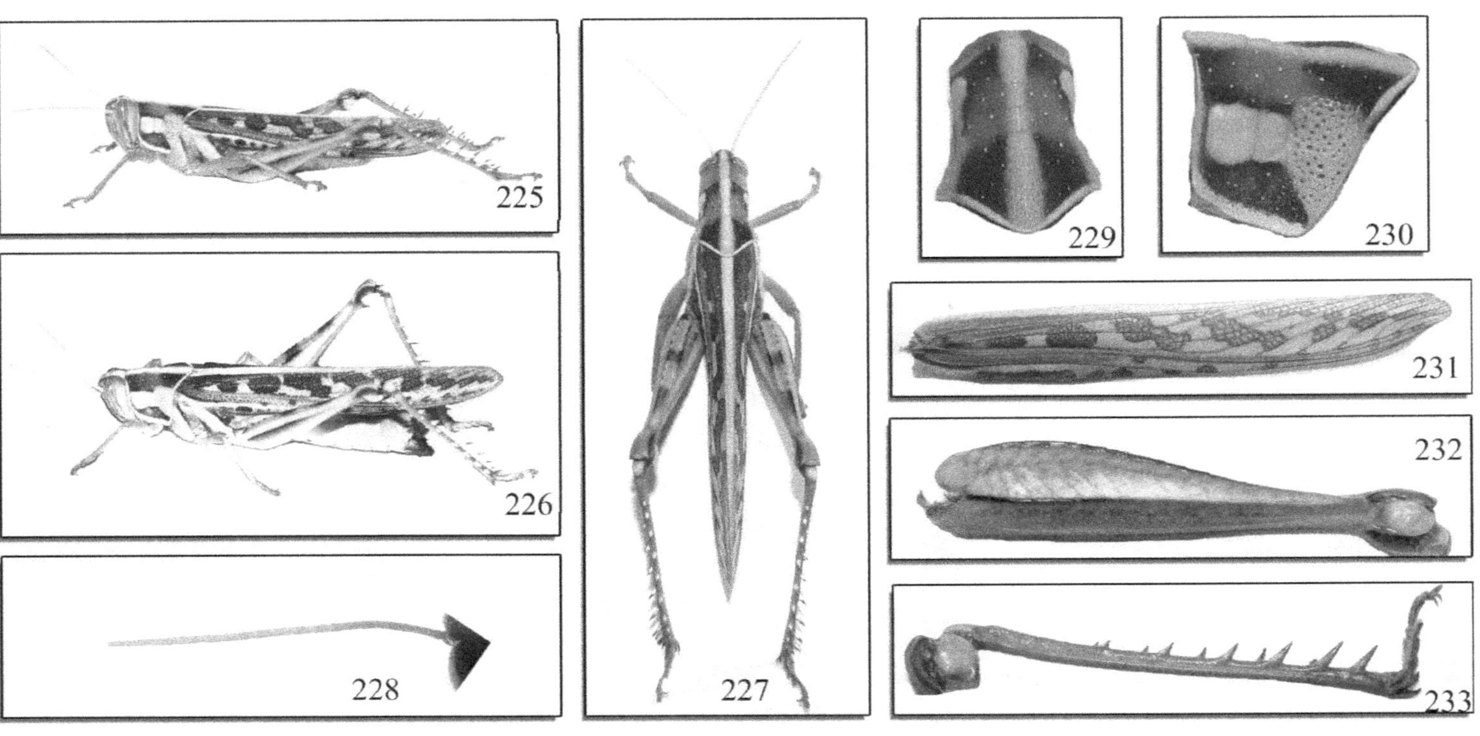

Plate 27: Figure 225: *Cyrtacanthacris tatarica* (Linn.) male; Figure 226: *C. tatarica* (Female); Figure 227: *C. tatarica* dorsal view; Figure 228: Antenna; Figure 229: Pronotum dorsal view; Figure 230: Pronotum lateral view; Figure 231: Tegmina; Figure 232: Femur; Figure 233: Tibia

Plate 28: Figure 234: *Anacridium flaviscence* (Fabrisius) female; Figure 235: *A. flaviscence* dorsal view; Figure 236: Lateral view of pronotum; Figure 237: Dorsal view of pronotum; Figure 238: Antenna; Figure 239: Tegmina; Figure 240: Femur; Figure 241: Tibia

Family–Acrididae

Subfamily–Romalienae

***Teratodes monticollis* Grey, 1832 (Plate 29, Figures 242 to 249)**

Green coloured, highly camouflagic. It is easily identified by their special type of pronotum. Head green; eyes brown, oval; antennae brown; pronotum (Figure 246) modified like hood and they extend above the head at front side and back side near about the center of tegmina, middle carina sharp, toothed at back side; tegmina (Figure 247) universally green and thick; femora (Figure 248) rough, toothed spines present on dorsal, lateral and ventral carina of femora; hind tibia (Figure 249) green, 9-10 red tipped spines present on lateral side. In male, head 9 mm long; pronotum 26 mm long; meso and meta thorax 11mm long; abdomen 22 mm long; tegmina 30 mm long; wing span 64 mm; body length 41mm long. It is reported only in the hilly region of the study area.

They feed on ber, badam, arjun and ain plants.

Family–Acrididae

Subfamily–Catantopinae

***Catantops pinguis* Stal. 1860 (Plate 30, Figures 250 to 258)**

Antennae brownish, brown towards the tip and pale towards the base; frontal ridge impressed in middle, starting from median ocellus up to clypeus; Tegmina (Figure 256) longer than the abdomen, finely mottled with brown; eyes brown oval; a faint black patch present on dorsolateral side of pronotum; whitish oblique band present on lateral side of meso and meta thorax. Pronotum (Figures 254 and 255) subcylindrical, slightly narrowing, prosternal tubercle thick, cylindrical or slightly anterior-posteriorly compressed with rounded apex, hind femora (Figure 257) stout, internal part reddish, two black bands present on femora. A single dark black spot present at the base of hind femora; tibia (Figure 258) red, 10 black tipped spines present on lateral side. In female, head 5 mm long; antennae 11.8 mm long, scape 1.7mm long and 1.5 mm wide, pedicel 0.9 mm long and 1 mm wide, flagella 8.7 mm long; pronotum 5 mm long; meso and meta thorax 9 mm long; abdomen 15mm long; tegmina 29 mm long; wing span 62 mm. In male, antennae 9 mm long; head 5mm long; prothorax 2 mm; pronotum 5 mm; meso and meta thorax 6 mm long;

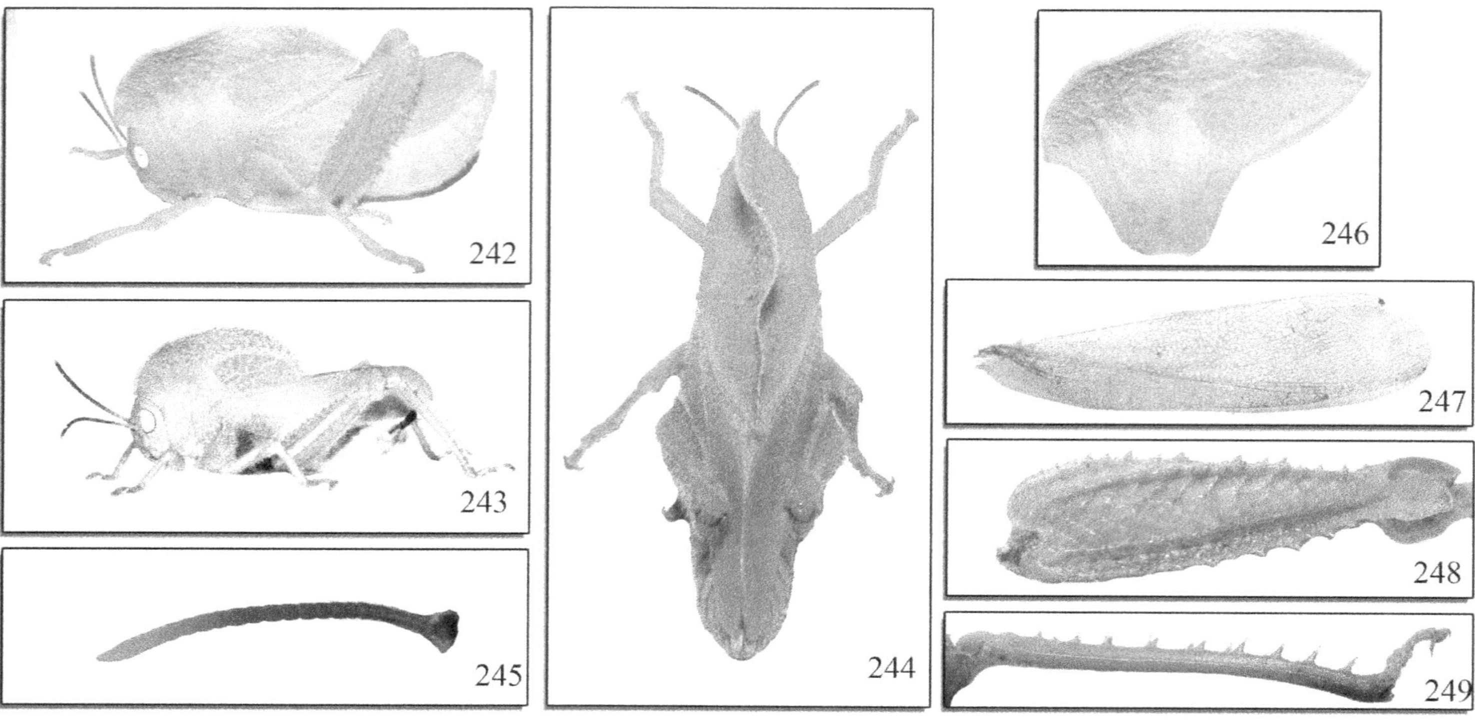

Plate 29: Figure 242: *Teratodes monticollis* Gray (Female); Figure 243: *T. monticollis* (male); Figure 244: Dosal view of *T. monticollis;* Figure 245: Antenna; Figure 246: Lateral view of pronotum; Figure 247: Tegmina; Figure 248: Femur; Figure 249: Tibia

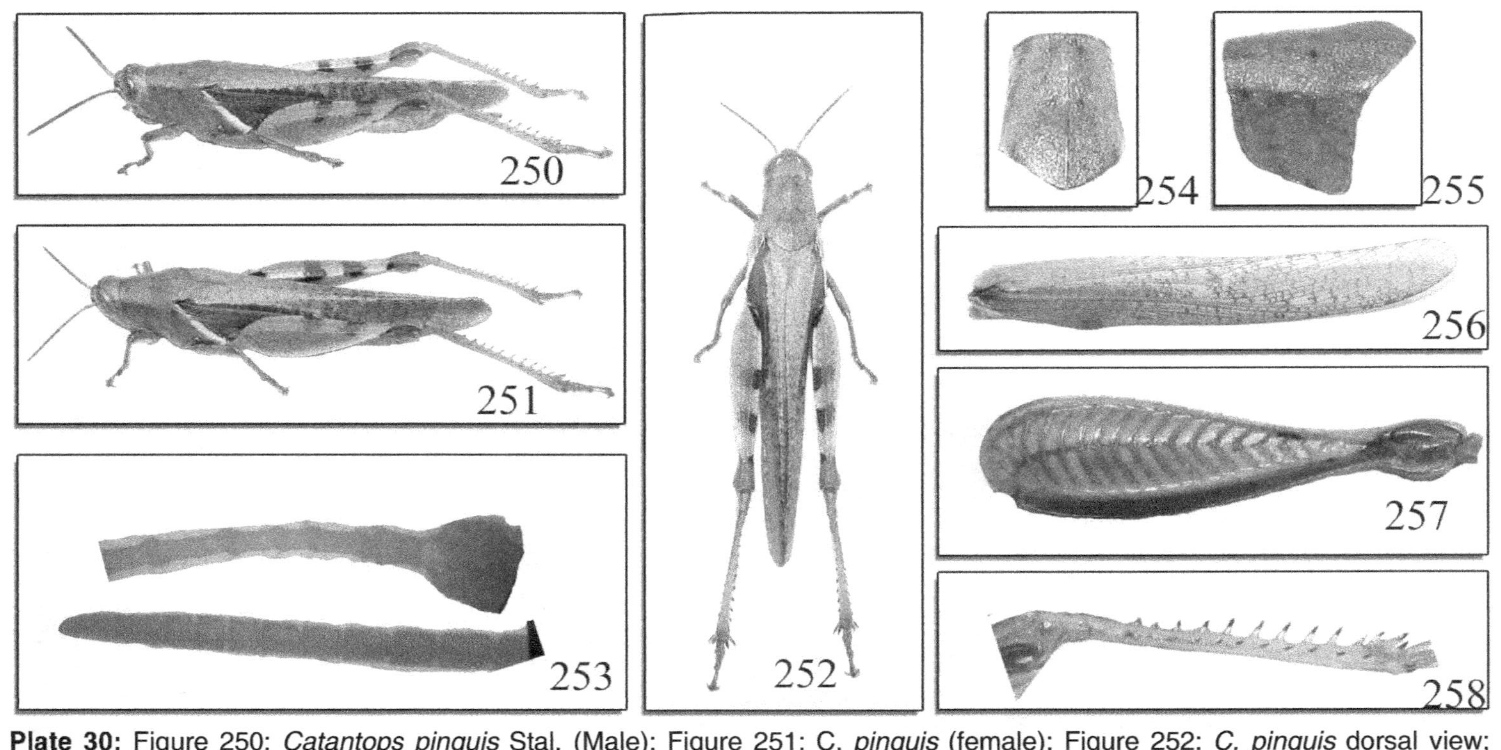

Plate 30: Figure 250: *Catantops pinguis* Stal. (Male); Figure 251: C. *pinguis* (female); Figure 252: *C. pinguis* dorsal view; Figure 253: Antenna (a–Posterior side, b–Anterior side); Figure 254: Pronotum dorsal view; Figure 255: Pronotum lateral view; Figure 256: Tegmina; Figure 257: Femur; Figure 258: Tibia

abdomen 14 mm long; tegmina 24 mm long; wing span 30 mm. They feed on plants like wheat, maize, paddy, tomato, etc.

Flagellar formula = 1L/W = 0.9, 4L/W = 1, 7L/W = 1.1, A = 1

Family–Acrididae

Subfamily–Catantopinae

***Xenocatantops humalis humalis* (Serville, 1839) (Plate 31, Figures 259 to 267)**

Antennae brownish, brown towards the tip and pale towards the base; eyes oval; pronotum (Figure 263,264) not subcylindrical, constricted. Prosternal tubercle conical, brown band present on lateral side of pronotum, below this band, a white narrow band present; whitish yellow band present on the lateral side of meso and meta thorax in oblique position; prosternal tubercle conical; hind femora (Figure 266) stout, ventral side of femora brown, two black bands present on dorsolateral side of hind femora and one black patch present on internal side at the base of femora; hind tibia (Figure 267) pale at the base and red towards the tip. In female; head 5 mm long; antennae 10.2 mm long, scape 1.2 mm long and 0.9 mm wide, pedicel 1 mm long and 0.8 mm wide, flagella 8 mm long; prothorax 2 mm long; meso and meta thorax 6 mm long; abdomen 12 mm long; body length 22 mm long; tegmina 21 mm long; wing span 44 mm long. It feeds on the crops like lady finger, tomato, fenugreek and found associated with semi dried grasses.

Fllagellar Formula – 1L/W =1.4, 4L/W =1.4, 7L/W = 1.4, A= 1.

Family–Acrididae

Subfamily–Catantopinae

***Stinocatantops splendens* (Thunberg, 1815) (Plate 32, Figures 268 to 276)**

Reddish brown; morphologically closer to the *Catantops pinguis*; antennae brown, pale at the base; eyes oval brown; head and pronotum (273,273) unicolour; Pronotum flattened, porsternal tubercle laterally compressed, median carina of pronotum smooth; oblique white band present on meso and metathorasic segment; tegmina (Figure 274) long, narrow with round apex; femora stout, three black patches present on dorsolateral region of hind femora; hind tibia (Figure 276) yellowish with 9-10 black tipped spines present. In female, head 6 mm long; antennae

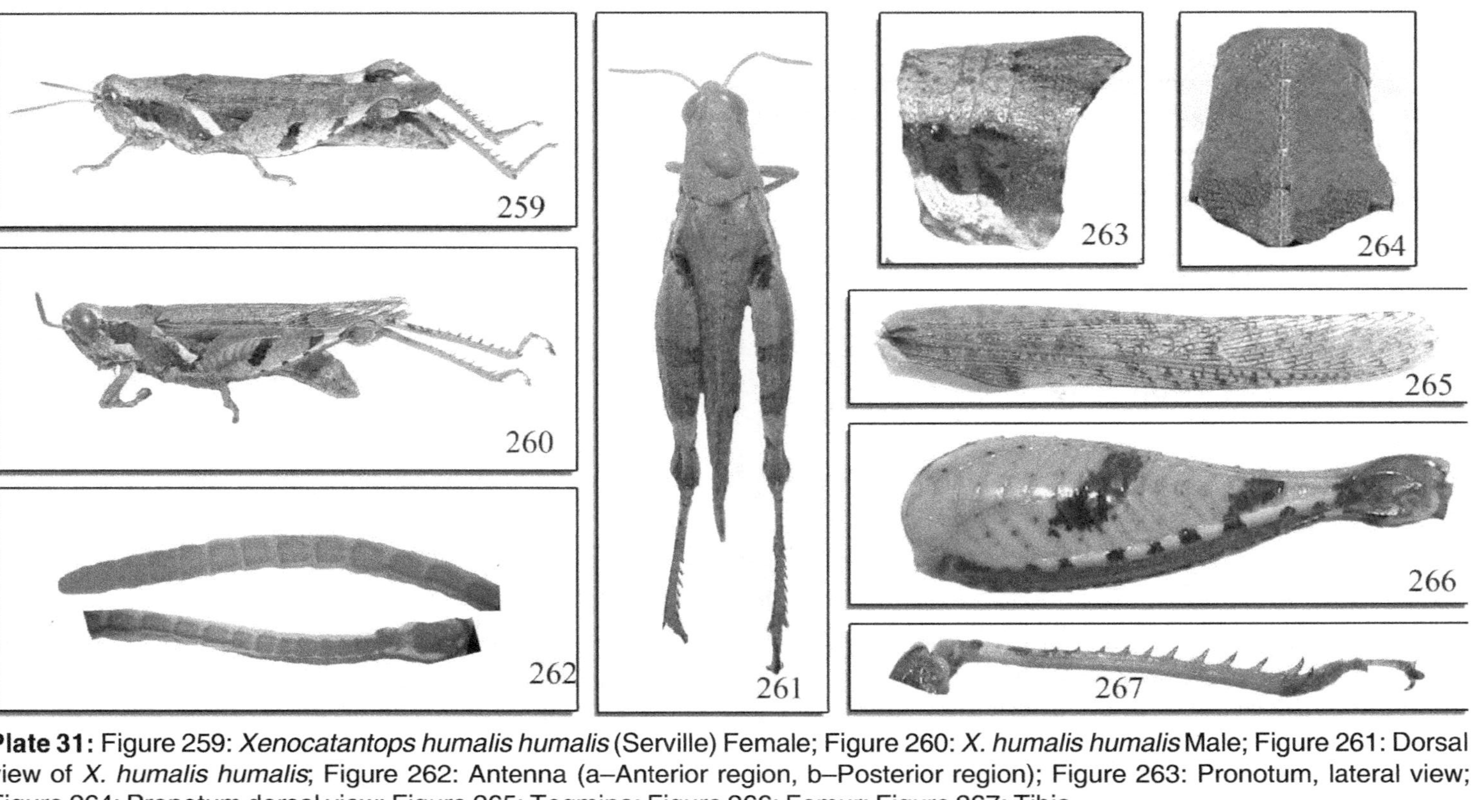

Plate 31: Figure 259: *Xenocatantops humalis humalis* (Serville) Female; Figure 260: *X. humalis humalis* Male; Figure 261: Dorsal view of *X. humalis humalis*; Figure 262: Antenna (a–Anterior region, b–Posterior region); Figure 263: Pronotum, lateral view; Figure 264: Pronotum dorsal view; Figure 265: Tegmina; Figure 266: Femur; Figure 267: Tibia

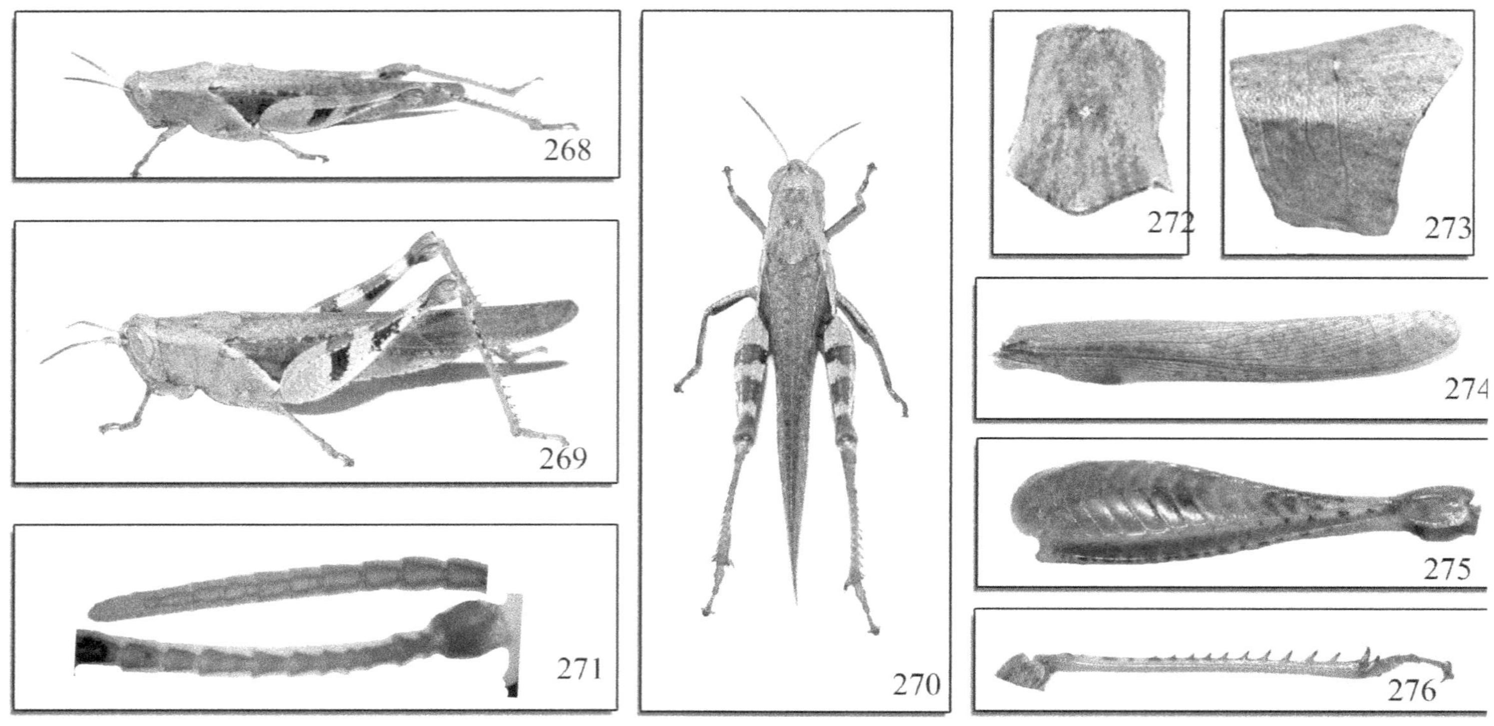

Plate 32: Figure 268: *Stenocatantops splendens* Female (Thunburg); Figure 269: *S. splendens* male; Figure 270: *S. splendens* dorsal view; Figure 271: Antemia (a–Anterior side, b–Posterior side); Figure 272: Pronotum dorsal view; Figure 273: Pronotum lateral view; Figure 274: Tegmina; Figure 275: Femur; Figure 276: Tibia

10.2 mm, scape 1.7 mm long and 1.5 mm wide, pedicel 1 mm long and 0.8 mm wide, flagella 7.5 mm long; pronotum 7 mm long; meso and meta thorax 8 mm long; abdomen 17 mm long; tegmina 29 mm; wing span 62 mm and in male, antennae 8 mm long; head 4 mm long; meso and meta thorax 6 mm long; abdomen 14 mm long; body length 25 mm; tegmina 24 mm; wing span 30 mm.

This species feed on wheat, pea plant, gliricidia and grasses.

Flagellar formula: 1L/W = 2.25, 4L/W = 1.75, 7L/W = 1.4, A = 1.8

Family–Acrididae

Subfamily–Eprypocnemidinae

Choroedocus robustus **(Serville, 1839) (Plate 33, Figures 277 to 284)**

Head broad; frontal ridge reddish, flat; antennae brown towards the tip and yellowish towards the base; eyes oval, brown; fastigium of vertex short, about as long as wide shallowly concave, obtusely rounded in front; pronotum (279, 280) carinated, median carina laterally compressed, a black band runs over the head and pronotum up to the posterior tip of pronotum; fore and mid legs brown, hind legs femora (Figure 283) reddish, pale ring present before the extremity. Dorsal side of femora decorated with black patch. Tegmina (Figure 283) decorated with black brown spots, anal area green. Subgenital plate conical, compressed towards apical region, and almost pointed at apex. In female, antennae 16 mm long; pronotum 9 mm long; thorax 16 mm long; abdomen 36 mm long; head 9 mm long (including mandible); tegmina 41 mm long; wing span 82 mm. Feed on young sugarcane, jowar, maize and grasses.

Family–Acrididae

Subfamily–Eprypocnemidinae

Tribe–Eyprepocenemidini

Eyprepocnemis alacris alacris **(Serville, 1838) (Plate 34, Figures 285 to 293)**

Medium size, brown colour; antennae pale; thick brown strip present on dorsal side of pronotum (Figure 289,290), median carina slightly carinated; tegmina (Figure 291) brown, dark brown spots present on tegmina; longitudinal black strip present on the outer side of hind femora.

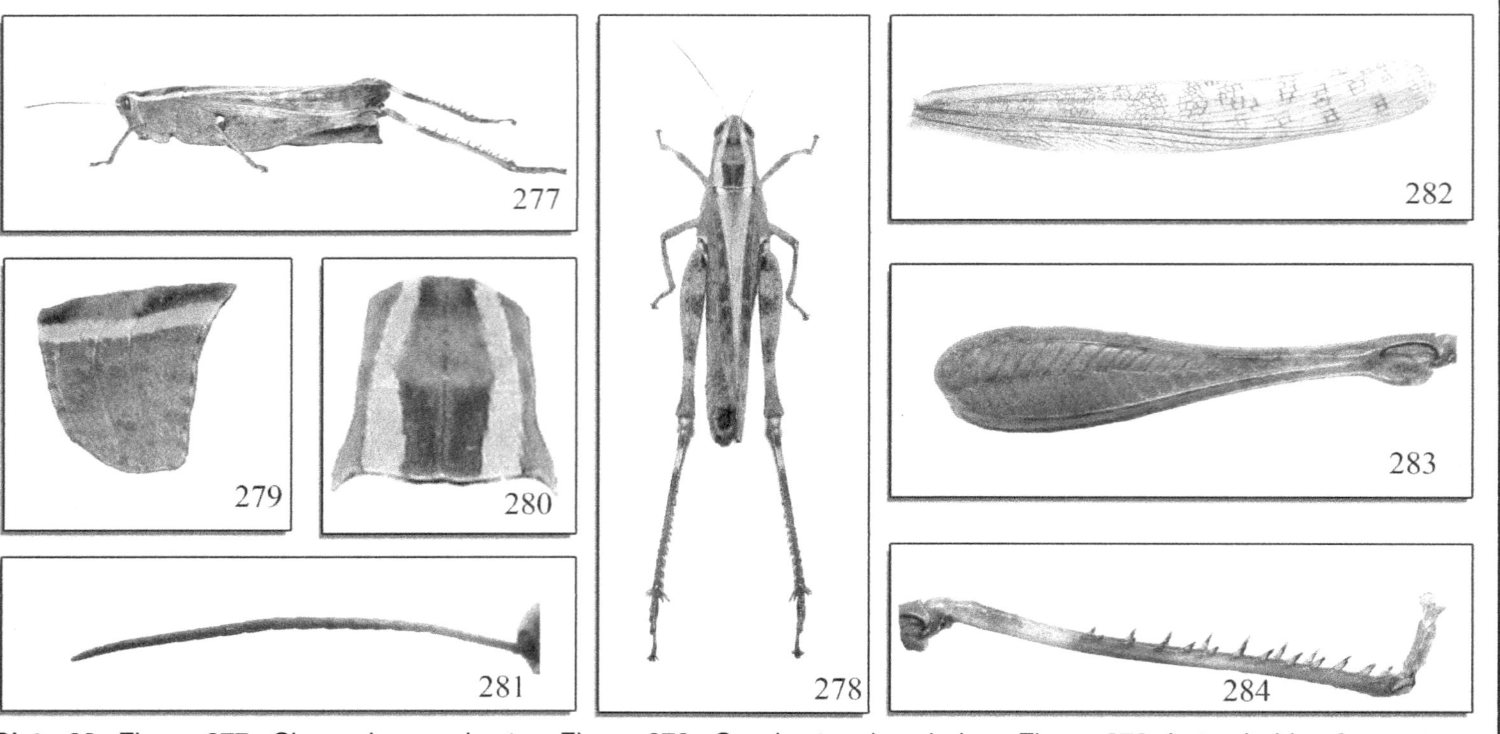

Plate 33: Figure 277: *Choroedocus robustus*; Figure 278: *C. robustus* dorsal view; Figure 279: Lateral side of pronotum; Figure 280: Dorsal view of pronotun; Figure 281: Antenna; Figure 282: Tegmina; Figure 283: Femur; Figure 284: Tibia

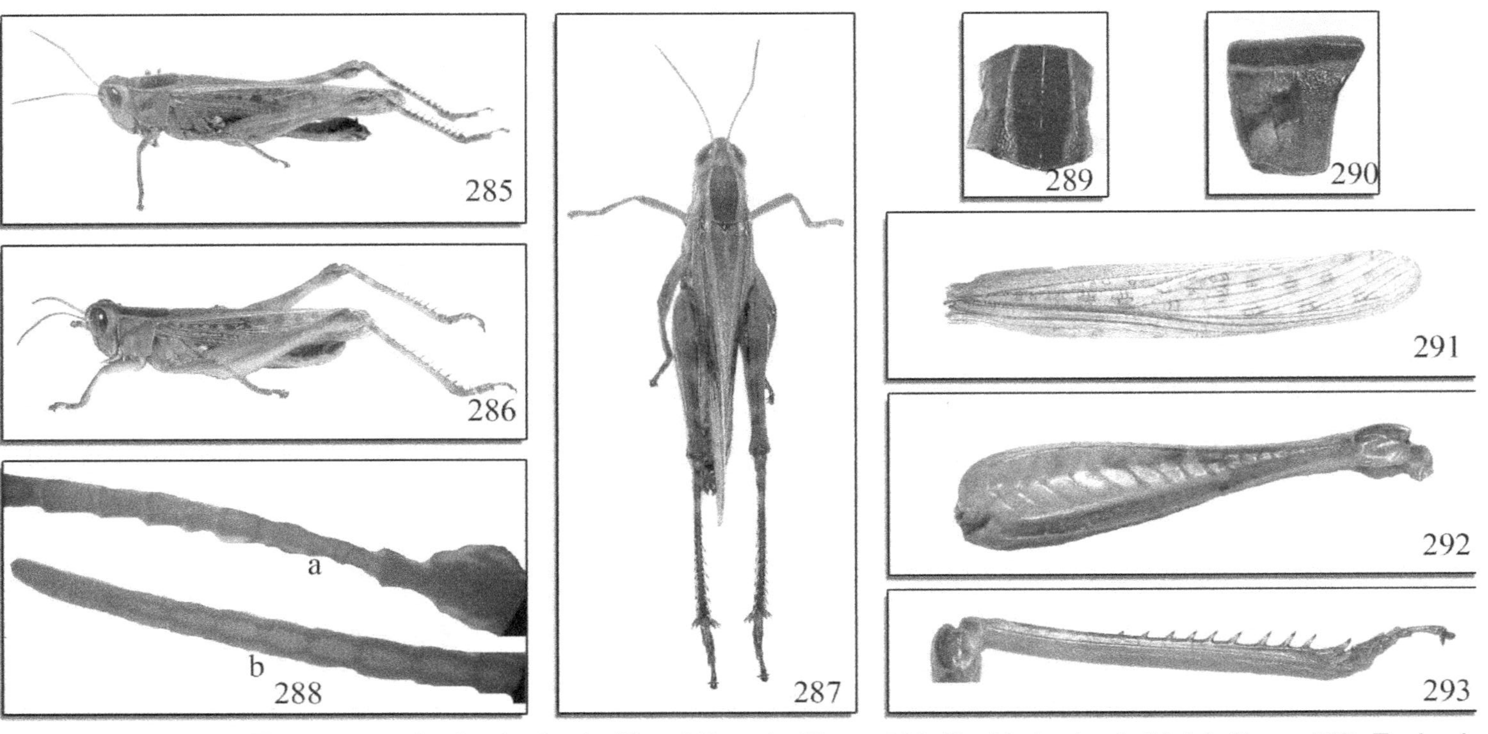

Plate 34: Figure 285: *Eyprepocnemis alacris alacris*, (Serv.) Female; Figure 286: *E. alacris alacris* (Male); Figure 287: *E. alacris alacris* dorsal view; Figure 288: Antenna (a–Posterior side, b–Anterior side); Figure 289: Pronotum dorsal view; Figure 290: Pronotum lateral view; Figure 291: Tegmina; Figure 292: Femur

The hind femur (Figure 292) moderately long, moderately produced beyond abdomen, not inflated basally, not strongly narrowing on apical half; Posterior tibia (Figure 293) with spine. Prosternal process almost spathulate, cylindrical with rounded apex, lateral side pale; tibia greenish, pale coloured two rings present at the base. In female, head 7 mm long; antennae 12.5 mm, scape 2 mm long and 1.3 mm wide, pedicel 1 mm long and 0.9 mm wide, flagella 9.5 mm long; prothorax 4 mm long; meso and meta thorax 7 mm long; abdomen 15 mm long; tegmina 25 mm long; wing span 54 mm. In male, antennae 8 mm long; head 5 mm long; pronotum 5 mm long; meso and meta thorax 6 mm long; abdomen 11 mm long; tegmina 19 mm long; wing span 40 mm. Feed on brinjal, paddy, jowar, pea, green gram, etc.

Flagellar formula: 1L/W = 2.2, 4L/W = 2.2, 7L/W = 2.25, A = 2.21

Family–Acrididae

Subfamily–Eprypocnemidinae

***Tylotropidius varicornis* (Walk, 1853) (Plate 35, Figures 294 to 302)**

Head brownish; eyes triangular; antennae brownish towards the tip; on mid dorsal region of pronotum present brownish strips (Figures 298,299); lateral carina pale, tegmina brown with row of whitish spot and pale longitudinal strips present on costal area; hind femora (Figure 301) broad at the base and narrow towards the tip; hind tibia (Figure 302) greenish brown with 12-14 black lateral spines. Subgenital plate navicular with somewhat pointed apex; circus a little compressed, apically slightly downcurved, with subacute apex. In female, antennae 15.5 mm, scape 1.5 mm long and 1.8 mm wide, pedicel 1mm long and 0.9 mm wide, flagella 13 mm long; head 8 mm long; prothorax 4 mm long, meso and meta thorax 9 mm long; abdomen 20 mm long; body length 38 mm long; tegmina 32 mm long; wing span 66 mm. In male, antennae 8 mm long; head 7mm long; prothorax 3 mm long; pronotum 5 mm long; meso and meta thorax 6 mm long; abdomen 16 mm long; tegmina 25 mm long; wing span 52 mm. Found associated with long dried grasses.

Flagellar formula: 1L/W = 2, 4L/W = 1.6, 7L/W = 1, A = 1.53

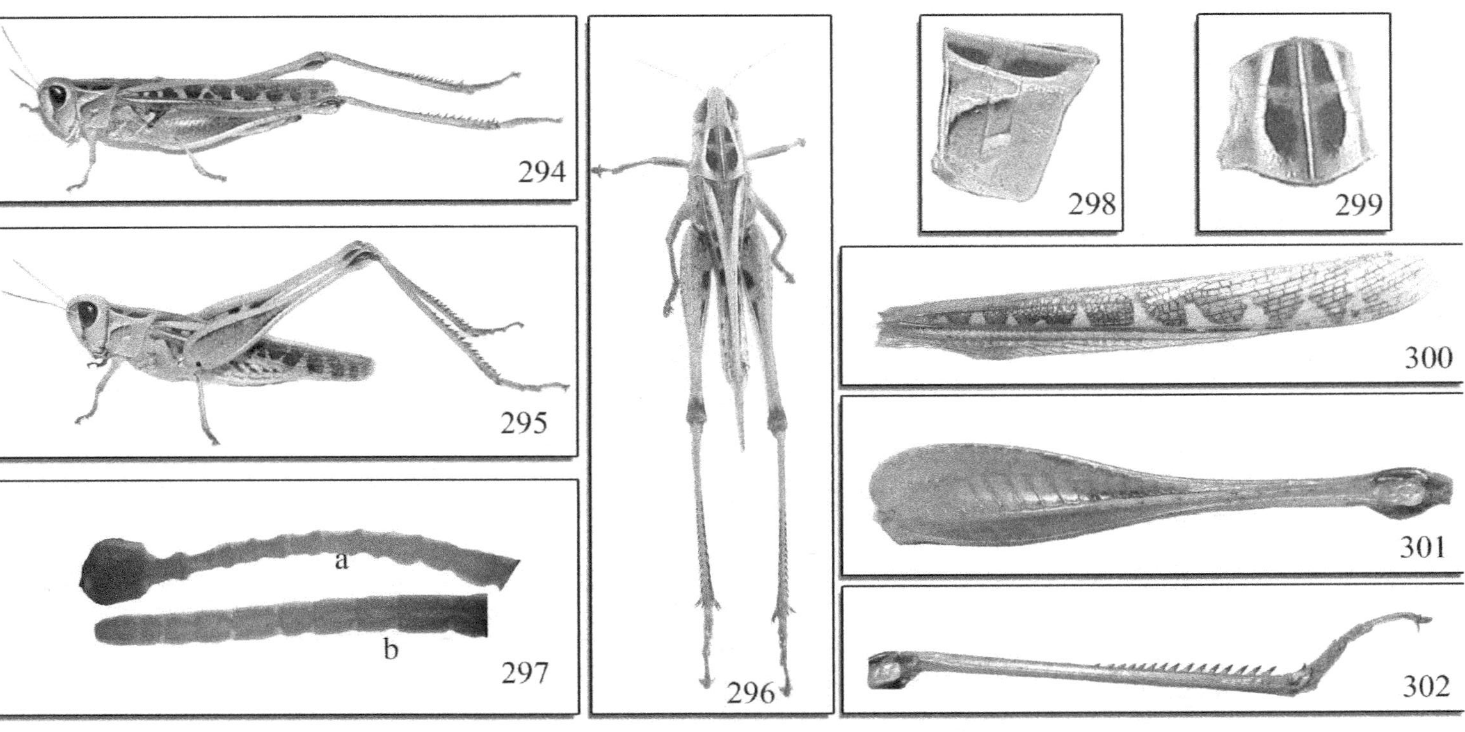

Plate 35: Figure 294: *Tyiotropidius varicornis* (Walk) female; Figure 295: *T. varicornis* (Male); Figure 296: *T. varicornis* dorsal view; Figure 297: Antenna (a–Posterior side, b–Anterior side); Figure 298: Lateral view of pronotum; Figure 299: Dorsal view of pronotum; Figure 300: Tegmina; Figure 301: Femur; Figure 302: Tibia

Key to the Genera of Family Pyrgomorphidae

1. Anterior margin of prosternum strongly reflexed and dilated; body short and stout; pronotum heavily rugosed and much widened behind; posterior femora not stout .... *Chrotogonus* Serville
- Anterior margi of prosternum neither reflexed nor dileted; body slender; pronotum not rugosed and not much widened behind; posterior femora not stout 2
2. Antennae remote for the eyes, placed in front of ocelli .... *Atractomorpha* Fabricius
- Antennae placed near the eyes and placed below the ocelli; wings absent, body with "Humped" dorsally; tegmina extending to the end of I[st] abdominal segment .... *Colemania* Bolivar
3. Pronotum roundly truncate in front, oubtusly angulated or rounded behind .... *Tagsta* Bol.

Key to the Species of Genus Chrotogonus Serville, 1893

1. Medium sized; hind wing hyline or occasionally tinged yellowish brown but never infumated .... *trachypterus* (Blanchard)
- Small sized; hind wing always infumated and extend nearly two thired as long as tegmina .... *oxypterus* (Blanchard)

Key to the Species of Genus *Atractomorpha* Sauss.

1. Pronotum pale above, granulated and impress-punctet on the hinder area; tegmina very long and pointed; ...... *blachirdi* Kirb
- Tegmina pointed, but not much longer than the wings 2
2. Pronotum slightly sinuated in front and obtusely angulated behind; hind margin of lateral lobes obtusely angulated .... *himalyica* Bol.
- Pronotum rather shorter than the head; hind margin of the lateral lobes of the pronotum deeply concave 3
3. Tegmina extending for one third of their length beyond the hind femora .... *psittacina* De Hann.

- Tegmina extending for one forth of their length beyond the hind femora 4

4. Frontal ridge between the antennae not sulcated; very short at tip of arolium .... *burry* Bol.

- Frontal ridge sulcated, or shortly compressed and arched between the antennae 5

5. Pronotum punctured and sparingly granulated, sub truncate in front and abtusely angulated behind; tegmina green; wings red at base .... *crenulata* Fab.

- Tegmina brown, motled with blakish; wings hyline........... *scabra* Thb.

Key to the Species of Genus *Tagsta* Bol.

1. Pronotum uniformly impress- punctuate; Tegmina with a black spot at base, containing an orange one; hind tibia red .... *notata* Brunn.

- Pronotum smooth concave at front and traingulae at base; tegmina faint grey coloured; hind tibia coloured .... *apludi* sp nov.

Family–Pyrgomorphidae

Subfamily–Orthacridinae

***Colemania sphenarioides* Bolivar, 1990 (Plate 36 Figure 303)**

Body greenish brown with reddish tinge; head conical; antennae (Figure 306) placed near the eyes and below the ocelli; antennae narrow towards the tip, blackish, pale at the base; vertex pointed, frontal side of face brown; eyes semi round; a dark brown band run over the lateral side of vertex to second abdominal segment. Pronotum (Figure 307, 308) broad; in female a black spot present on dorsal side of pronotum; prosternum accutly tuberculated; supra annal plate forming a long tringle; meso thorax broad; tegmina (Figure 309) short, extend to the end of Ist abdominal segment, wings absent; a black band present on mid lateral region of abdomen; a yellow band runs from the base of antennae to hind coxa on lateral side; legs greenish yellow; tarsus pinkish; Body humped. In female; head 11 mm long; antennae 17.2 mm, scape 2.7 mm long and 1.3 mm wide, pedicel 1mm long and 1.1 mm wide, flagella 13.5 mm long; pronotum

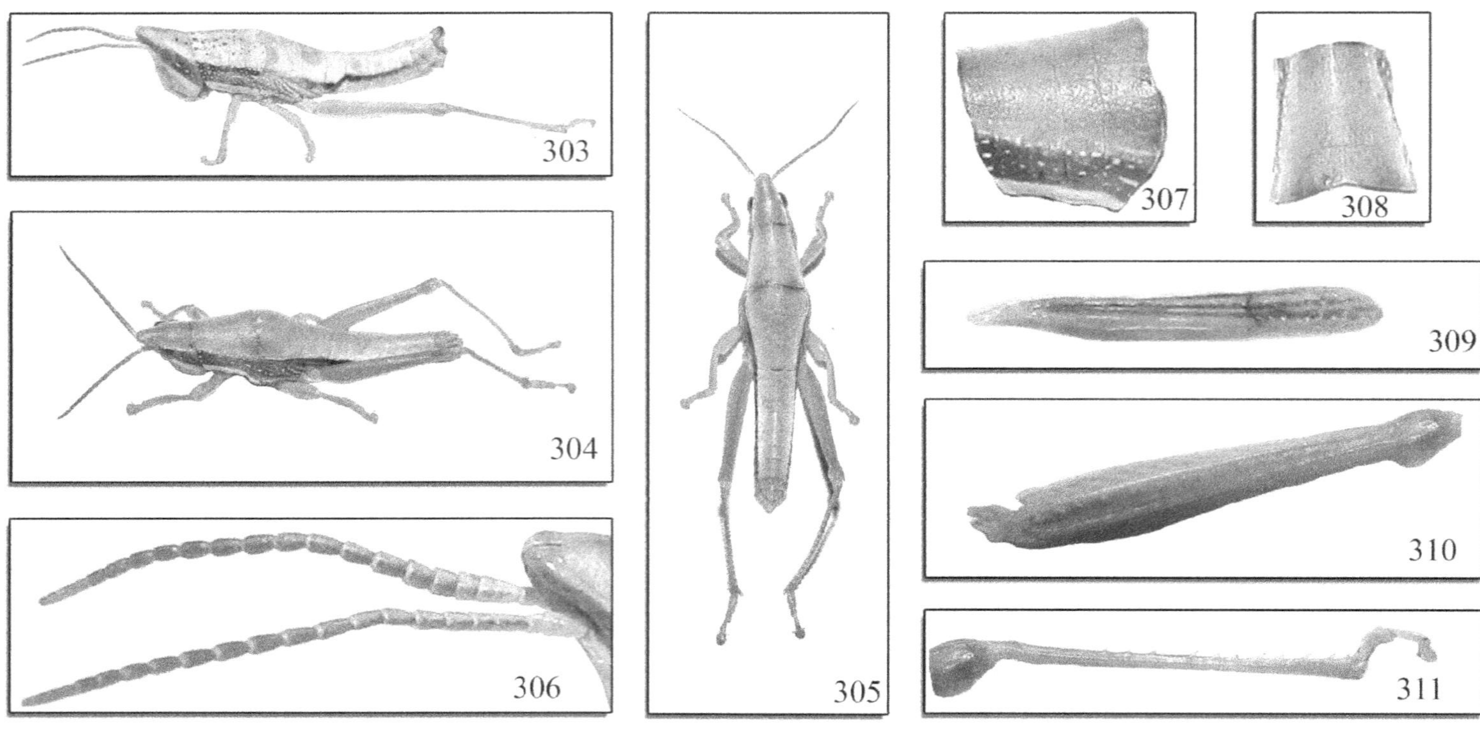

Plate 36: Figure 303: *Colemania sphenarioides* Bolivar (Female); Figure 304: *C. sphenarioides* (male); Figure 305: *C. sphenarioides* dorsal view; Figure 306: Antennae; Figure 307: Lateral view of pronotum; Figure 308: Dorsal view of pronotum; Figure 309: Tegmina; Figure 310: Femur; Figure 311: Tibia

7 mm; meso and meta thorax 6 mm long; abdomen 15 mm long; tegmina 8 mm long; hind femora 14 mm long. In male, antennae 13 mm long; head 11mm long; pronotum 6 mm long; meso and meta thorax 6 mm long; abdomen 15 mm long. Feed on long grasses in hilly region.

Fllagellar formula – 1L/W = 1.76, 4L/W = 1.30 7L/W – 1.9, A= 1.65

Family–Pyrgomorphidae

Subfamily–Pyrgomorphinae

***Chrotogonus oxypterus* Blanchard, 1836 (Plate 37, Figures 312 to 320)**

It is dorsoventraly flattened, small brown coloured; eyes round; antennae (Figure 315) brown; pronotum (Figures 316,317) heavily rugosed and much widened behind; median carina not differentiated; lateral carina sickle like; anterior margin of prosternum strongly refluxed and dilated; tegmina (Figure 318) rough and short; hind wing always infumated posterior; femora not stout, brown in colour, ventral side yellow with black spots; hind tibia (Figure 320) greenish brown with 7-8 black tipped spine present; body short and stout. In female; head 5 mm long; antennae 10.4 mm, scape 2.4 mm long and 1.4 mm wide, pedicel 1 mm long and 1 mm wide, flagella 7 mm long; pronotum 4.5 mm long; meso and meta thorax 5 mm long; abdomen 10 mm long; body length 20 mm long; tegmina 11 mm long. In male, antennae 5 mm long; head 3 mm long; pronotum 3 mm long; meso and meta thorax 3mm long; abdomen 6 mm long; tegmina 7 mm long. This species feed on surface vegetation and green gram, black gram, soyabean, pea, etc.

Flagellar formula = 1L/W = 1.62, 4L/W = 1.85, 7L/W = 1.25, A = 1.57

Family–Pyrgomorphidae

Subfamily–Pyrgomorphinae

***Atractomorpha crenulata* (Fabricius, 1793) (Plate 38, Figures 321 to 329)**

Body green coloured, head conical; eyes oval, brown; frontal ridge shortly compressed and arched between the antennae; antennae green, they remote for the eyes, placed in front of ocelli; median and lateral carina slightly shown; tegmina (Figure 327) green, pointed at the tip, two black or silvery spots present on costal area; tegmina extending for one forth of their length beyond the hind femora. The hind femora (Figure 328) narrow,

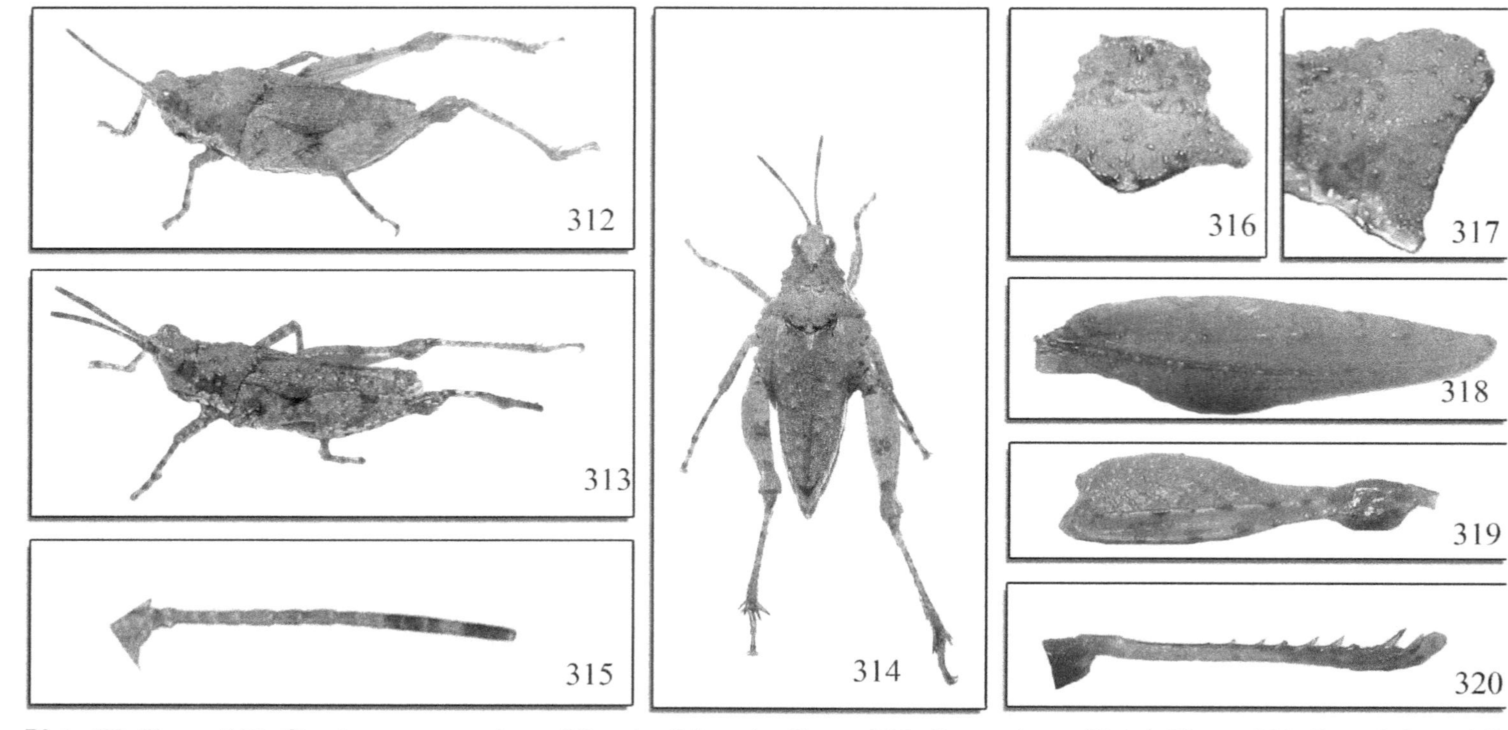

Plate 37: Figure 312: *Chrotogonus oxypterus* (Blanched) female; Figure 213: *C. oxypterus* (Male); Figure 314: Dorsal view of *C. oxypterus;* Figure 315: Antenna; Figure 316: Dorsal view of pronotum; Figure 317: Lateral view of pronotum; Figure 318: Tegmina; Figure 319: Femur; Figure 320: Tibia; Figure 329: Tibia

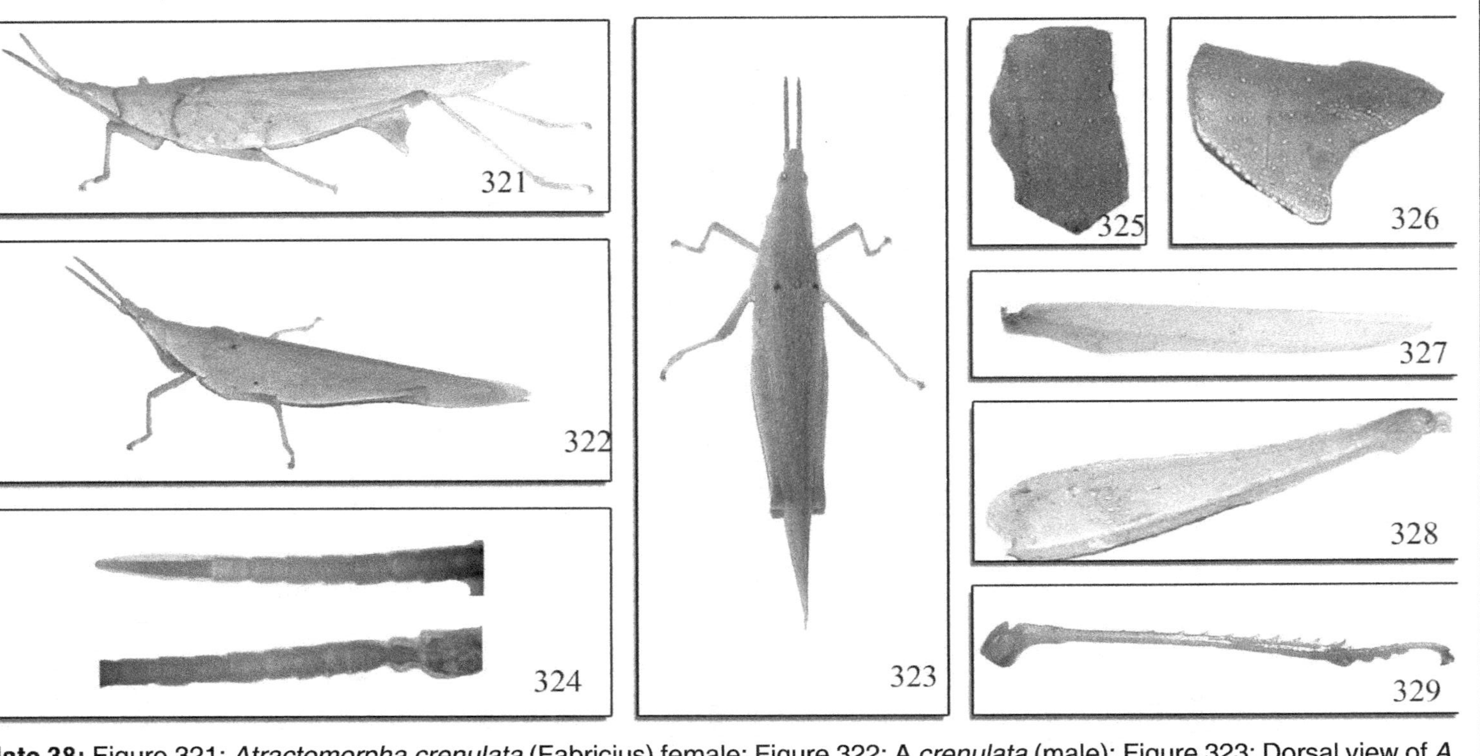

Plate 38: Figure 321: *Atractomorpha crenulata* (Fabricius) female; Figure 322: A *crenulata* (male); Figure 323: Dorsal view of *A. crenulata*; Figure 324: Antenna (a–Anterior side, b–Posterior side); Figure 325: Dorsal view of pronotum; Figure 326: Lateral view of pronotum; Figure 327: Tegmina; Figure 328: Femur

green; tibia (Figure 320) greenish, 9-10 spines present on lateral side of tibia. In female, head 6 mm long; antennae 15.2 mm, scape 1.8 mm long and 1.3 mm wide, pedicel 0.8 mm long and 1mm wide, flagella 12.6 mm long; prothorax 3 mm; meso and meta thorax 6 mm long; abdomen 13 mm long; tegmina 25 mm long; wing span 52 mm. In male, antennae 6 mm long; head 6 mm long; prothorax 2.5 mm; pronotum 3 mm long; meso and meta thorax 4 mm long; abdomen 7 mm long; tegmina 15 mm long; wing span 32 mm.

Fllagellar Formula –1L/W = 1.1, 4L/W = 0.9, 7L/W =1.37, A= 1.12

Family–Pyrgomorphidae

Subfamily–Pyrgomorphinae

The Genus–*Tagasta* Bolivar 1905

The genus *Tagsta* is raised by Boliver in 1905. It shows following characters:

1. Body subfusiform, slightly compressed, pubsent above.
2. Head conical, shorter than the pronotum or equal length.
3. Tempora widened in front, only separated by a short suture.
4. Front very oblique, frontal ridge much flattened, hardly sulkated,
5. Shortly compressed between the antennae.
6. Antennae filiform, inserted between the eyes, basal joint slightly flattened.
7. Eyes rounded, ocelli distinct, cheek granulated.
8. Pronotum pubsent, rounded truncate in front, obtusely angulated or rounded behind with the median carina very slightly indicated or obsolete, and the lateral carina obsolete.
9. The sulci slightly marked and the hind sulcus placed behind the middle.
10. The prozona considerably longer than the metazoan; the deflexed lobes distinctly higher behind, the lower margin oblique, subsequent, bordered with whitish.
11. Tegmina not or scarcerly longer than the hind femora, with the costal area considerably expanded near the base.
12. Wings distinctly shorter than the tegmina, red or hyaline.

13. Legs long and slender, front femora distinctly thickened in the male, hind femora compressed, outer area with radiating ridge.
14. Hind tibia with rounded spines and with an outer apical spine. above.
15. Prosternum strumose.
16. Metasternal lobes separated by a longer or shorter space.
17. Metasternal foveolae separated by a transverse space.
18. Valves of the ovipositor sinuated.

From India, two species have been reported under this genus.

TAGASTA APLUDI SP NOV (Plate 39, Figure 330 to 340)

Female (Figure 331)

23 mm long, grey colour, body slightly compressed.

Head

6 mm long and 2.5 mm wide, conical; dorsal side thick grey and ventral side whitish; eyes semi round; frontal ridge compressed; lateral carina slightly shown; a row of small yellow granules present at behind the eyes on lateral side of face; foveole well developed; medium carina at dorsal side of head, slightly pointed.

Antennae (Figure 334)

Short filiform, blackish towards the tip and yellowish towards the base, inserted between the cavity of foveolar and frontal ridge, basal segment broad and flate; scape–2 mm long, 1mm wide; pedicel 0.9 mm long, 0.6 mm wide, Flagellum = 12 mm long.

Flagellar formula: 1L/W=1.43, 4L/W = 1.09, 7L/W = 1.14, A=1.226

Thorax

6mm long and 2 mm wide; pronotum (Figures 8,9) smooth concave at front and triangular at base, 5 mm long and 2 mm wide; median carina slightly shown, lateral carina obliquely present, a single sulci present on lateral side of pronotum.

Fore wing (Tegmina) (Figure 333)

Yellowish grey, dark-grey at the tip 17 mm long and 2.5 mm wide.

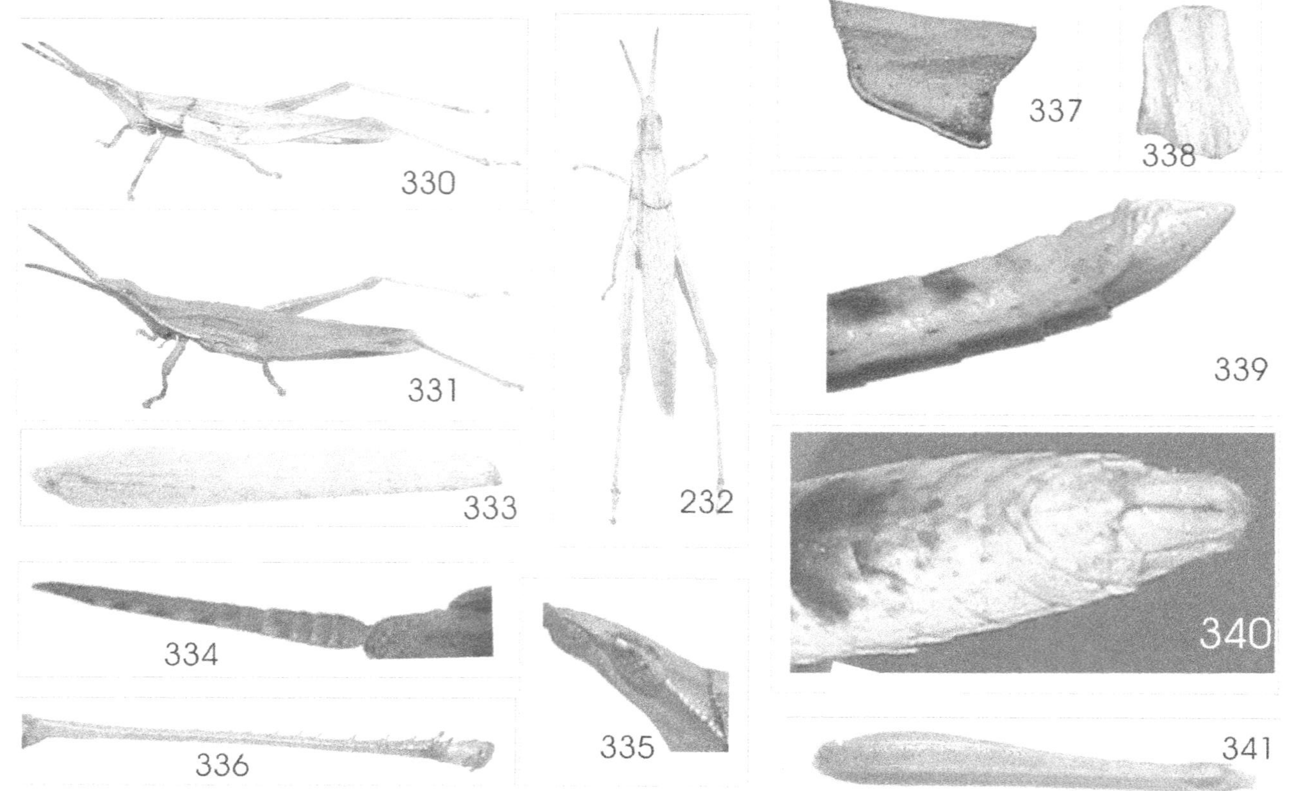

Plate 39: Figure 330: *Tagasta apludi* sp.nov. (Male); Figure 331: *T. apludi* (Female); Figure 332: *T. apludi* dorsal view; Figure 333: Tegmina; Figure 334: Antenna; Figure 335: Side view of head; Figure 336: Tibia; Figure 337: Lateral side of pronotum; Figure 338: Dorsal side of pronotum; Figure 339: Abdomenal tip of mak; Figure 340: Abdomenal tip of female; Figure 341: Femur

Hind Wing

Reddish towards the base and yellowish towards the tip, thin membranus, 8 mm wide and 17 mm long.

Fore Leg

Yellowish grey, 7.2 mm long; coxa 0.6 mm long; femur 3 mm long, smooth; tibia yellowish grey, a row of small point spine present at lateralside of tibia. Tibia 2 mm long, tarsus 1.6 mm long, first two segment short, at tip of 3rd segment present arolium and two pointed claws.

Mid Leg

Coxa whitish, 0.5 mm long; femur yellowish grey, 3.2 mm long; tibia 1.8 mm long; tarsus short, 1.2 mm long; as like fore leg first two tarsal segment short, 3rd segment long with claws and arolium.

Hind Leg

Yellowish grey; coxa whitish 1.2 mm long; femur (Figure 341) slightly compressed, irregular blackish small spot present over the femur, 10 mm long; tibia 8.5 mm long grey coloured, a row of nine black tiped spine present on lateral side of tibia, two pair of calcaria present on lateral side at tip of tibia; tarsus 2.3 mm long, arolium and hocked claws present at the tip of last tarsal segment.

Abdomen

Bluish patches present on dorsal side of abdomen, ventral side yellowish, first abdomenal segment submerged in the meta thorax; IInd abdomenal tergite 1.5 mm long; IIIrd tergite 2 mm long, IVth tergite 1.5 mm long, V^{th} abdomental tergite 1.2 mm long; VIth abdomenal tergite 1mm long, VIIth tergite 0.8 mm long, IXth tergite and X^{th} tergite modified as ovipositor (Figure 340), oviopositor short, 1.8 mm long, anal cerci 0.4 mm long.

Male (Figure 330)

Smaller than the female.

Head

5.6 mm long, 1.8 mm wide as like a female; eyes semiround, grey coloured othere character more or less similar to the female.

Antennae

6.7 mm long, Filiform, greyish brown coloured. Scape- 1.1 mm long, 0.9 mm wide; pedicel 0.6 mm long, 0.7 mm wide and flagellum 5mm long.

Flagellar formula- 1L/W= 0.75, 4L/W=0.87, 7L/W=1.2, A= 0.94.

Thorax

5.8 mm long and 2.8 mm wide; pronotum 3.4 mm long and 1.6 mm wide, median carina slightly shown, lateral carina oblique, two sulci present, one are complete and one are half; meso and metathorax 3.2 mm long and 2.3 mm wide.

Fore Wing

(Tegmina): 12 mm long and 1.2 mm wide, yellowish grey colour.

Hind Wing

Reddish, dark red towards the base and faint grey at the tip. 12 mm long and 6 mm wide.

Fore Leg

Faint grey coloured; coxa 0.4 mm long; femur 1.9 mm long; tibia 1.6 mm long; tarsae 1.2 mm long, arolium and hooked claws present at tip.

Mid Leg

Coxa short, femur 1.9mm long; tibia dark grey, 1.5 mm long; tarsae 1.2 mm long.

Hind Leg

Yellowish grey, delicate; coxa 0.7 mm long; femur 8 mm long; 9 black tiped spine present on lateral side of tibia and two pair of calcaria present at tip of tibia; tarsus 1.2 mm long with claws and aronium.

Abdomen

Blue patches present on dorsal side, first segment short submerged in metathorax, XIth segment modified as a genitalia, 1st abdomenal tergite 0.6 mm long IInd tergite 0.9 mm long, IIIrd tergite 1mm long, IVth 1.1 mm long, Vth 0.8 mm long, VIth 0.6 mm long, VIIth 0.5 mm long, genitalia (Figure 339) 1.6 mm long.

Holotype

Female, India, Maharashtra, Kolhapur, Coll, Bhusnar A.R. 22- III–2012. Head, antennae, legs, and wings mounted on the card sheet, rest body parts pinned and labelled as above. Male, India, Maharashtra, Kolhapur. Coll. Bhusnar A.R. 22 III 2012. Head, antennae, leg, wing and abdomen mounted on the card sheet, rest body parts pinned and labelled as above.

Paratype

6♀, 2♂, sex ratio (M.F.) 1:3, Coll. Bhusnar A.R. March 2012 to April 2012.

Etymology

The species described in the text is found mostly associated with Grass *Apluda mutica* hence the name *Tagasta apludi.*

Remarks

Tagasta apludi. sp nov runs close to *Tagasta indica* L. Bol. by having following characters:

1. Antennae inserted near the eyes.
2. Cheeks with row of large yellow granuls.
3. Pronotum rounded in front and obtusel angulated behind.
4. Wings deep rose colour, with costal area sub hyaline.

However, it differs from above species by having following features:

1. Body colour yellowish grey.
2. Wings length equal to the Tegmina.
3. Latteral carina middaly cut down, Posterior region of latteral carina run down towards the anterobasal region of pronotum.
4. Hind femoral and tibial colour and number of spine.
5. Body and Tegminal length praportion.
6. Tegmina not as long as the hind femora.
7. Flagellar formula:

   Female: 1L/W = 1.43, 4L/W = 1.09, 7L/W = 1.14 A=1.22

   Male: 1L/W= 0.75, 4L/W=0.87, 7L/W=1.2, A= 0.94.

Key to the Sub Familes of the Family TETRIGIDAE

1. Posterior angles of lateral lobes of pronotum acutely produced outwards and generally spined; first segment of posterior tarsi generally longer than the third..... SCELIMENINAE

– Posterior angles of the lateral lobes of pronotum rarely acutely spined; first segment of posterior tarsi generally not longer than the third 2

2. Posterior angles of the lateral lobes of pronotum little produced outwards, obliquely truncate behind, very rarely acutely spinose; first and third posterior tarsal segment nearly equal in length .... METRODORINAE

– Posterior angles of the lateral lobes of pronotum turned downwards, more or less rounded; first and third posterior tarsal segments unequal in length, first rather longer than the third .... TETRIGINAE

Key to the Genera of Subfamily Sclimeninae

1. Posterior tibiae and first tarsal segment expanded 2

– Posterior tibiae and first tarsal segment not expanded 3

2. Apical margin of the posterior tibia and first segment of posterior tarsi with lamellar expansion; spine of lateral lobes tarsi with wide lamellar expansion; spine of lateral lobes of pronotum well developed, apex not hooked; small tubercles present on the anterior margin of pronotum below the eyes, in front on the median carina of pronotum, on humeral angles and sometimes on the lateral carina of pronotum also; generally small tuberculated teeth present on the inferior margin of posterior femora ....

Scelimena Serv.

– Apical margin of the posterior tibia and the first segment of posterior tarsi more or less expanded but not laminated; spine of lateral lobes of pronotum weakly developed and hooked at apex; tubercle weakly developed on anterior margin below the eyes and in front the median carina of pronotum, but these are absent on humeral angales and lateral carinae of pronotum; inferior margin of posterior femora smooth .... *Indoscelinena* Gunther

3. Lateral margin of pronotum behind the shoulders, from the side view, distinctly upward and from top view bent inwards, wider between shoulders; generally brachypterous with strong sculpture; spine of posterior angles of lateral lobes of pronotum distinctly projects out posteriorly or transversely .... *Thoradonta* Hancock
– Lateral margin of pronotum behind the shoulders produced 4
4. Vertex broader than an eye or sometimes in females only just as broad as an eye but never narrower; eye not raised above the pronotum .... *Criotettix* Bol.
– Vretex broder than the eyes with slight median carina; frontal carina not prominent .... *Acanthalobus* Hanc.

Key to the Species of Genus of *Scelimena*

1. Lateral lobes of pronotum with only one spine 2
– Lateral lobes of pronotum with two sipnes, curving forwords .... *Logani* Hanc.
2. Spine nearly straight; lamellae of hind tibia very wide, sub higline .... *harpago,* serve.
– Spine distinctly curved forwards lamet llae of hind tibia and first joint of hind tarsi narrower, more gradually formed and only sub hyline at edges .... *gavialis* Sauss

Key to the Species of Genus of *Thoradonta* Honcock

1. Upper lobes of posterior angles of lobes of pronotum with small or medium sized spine, directed posteriorly front margin not constricted; upper marginal and paginal areas of posterior femora with a series of nodules ....*nodulosa* Stal.
– Upper lobes of posterior angles of lateral lobes of pronotum with medium sized spines, directed transversely, front margin constricted; upper marginal and paginal areas of posterior femora without nodules .... *purthi* Gunteri

Key to the Genera of Subfamily Metrodorinae

1. Vertex process, in profile, down up to about half the length of eyes in front, front margin of vertex, in top view, forms an obtuse angle, paired ocelli placed not always below the middle of eyes. Flight organs are always present .... *Spadotettix* Hancock

– Vertex process in profile not drawn out, paired ocelli placed not always below the middle of eyes. Flight organ present or absent 2

2. Head exerimated above the pronotal surface; big and globular eyes approxte and raised above the pronotum .... *Systolederus* Boliver

– Head not exerted above the pronotal surface; eyes bigger and approximate, especially at the anterior and thus making the vertex triangular; pronotum not ascended or rarely in females weakly developed .... *Teredorus* Hancock

Key to the Species of Genus *Systolederus* Bolivar

1. Frist two pulvilli of the hind tarsi of equal length, the third longer .... *cinereus*, Brunn.

– All three pulvilai of equal length; process of pronotum produced considerably beyond the hind femora .... *greeni* Bol.

Key to the Species of Genus *Acanthalobus* Hanc.

1. Wings approximately as long as the pronotum 2

– Wings more or less abbreviated; Pronotum differs in being more rugose and more cuneate posteriorly .... *cuneatus*, Han

2. Pronotum granulated, but not rugose; lateral spine acute .... *flavopictus*, Bol.

– Pronotum rugose or tuberculated, deepley punctate; lateral spine derected outward or slihgtly bacwords ...... *miliarius* Bol.

Key to the Genera of Subfamily Tetriginae

1. vertex, viewed in profile, not at all produced in front of eyes 2

– Vertex, viewed in profile, produced before the eyes, angulate, viewed from above, wider than an eye, not arrowed in front .... *Tetrix* Latreille

2. Vertex and eyes obiously raised above the level of pronotum, antennae inserted between the lower border of eyes 3

- Vertex and eyes very little or not at all raised above the level of the pronotum, antennae scarcely placed below the middle of eyes 4

3. Vertex narrower than an eyes and truncate, frontal costa arcuately produced between the middle of the eyes, antennae inserted below the middle of eyes, median carina of pronotum not undulated, posterior femora with less conspicuous nodosities on the external pagina .... *Euparatettix* Hancock

- Vertex still narrower than an eye, antennae inserted distinctly below the inferior 'margin of eyes, median carina of pronotum a little or very well undulated, posterior femora with distinct nodosities on external pagina .... *Ergatettix* Kirby

4. Body generally smooth or a little granulose, vertex equal to or a little narrower than an eye, generally expanded, subtransverse or triangularly carinated in front, abbreviated carinae distinctly present, median carina of pronotum percurrent, not at all interrupted .... *Hedotettix* Bolivar

Key to the Species of Genus *Euparatetix* Hancok, 1904

- Hind tibia dense black with white annulations just behind the knees; frontal costa generally narrowly sulcate ......*personatus* Bol.

- Hind tibia subunicoloured or obscurely marked but never dense black; frontal costa generally narrowly sulcate 2

Stature very slender; head exerted; pronotum narrow between shoulders; median carina of pronotum substraight percurrent .... *tenuis* Hancock

- Status more robust and larger; head a littele exerted; pronotum moderately dilated between shoulders; median carina of pronotum arcute forword, often little undulate before the shoulders, low on shoulder and straight behind .... *histricus* (Stal)

Key to the Species of Genus *Hedotetix* Bolivar, 1887

1. Frontal carina widely sulcate, vertex broad as an eyes; wings longer than the pronotum .... *gracilis* D. Haan
- Upper surface is brown, with several raised longitudinal whitish lines .... *linefera* Walk.

Key to the Species of Genus *Ergateteix* Kirbay

1 Median carina of pronotum distinctly undulate, lateral carina with small dilated lobes on the hind process of pronotum; median femora sufficiently flattened, margins lobate; hind femora strongly tuberculate on external pagina .... *guntheri* Steinmann
- Median carina of pornotum indistinctly or distinctly undulate but not like the above; lateral carinae without dilated lobes on the hind process of pronotum; median femora less flattened, margins indistinctly or a little lobate; hind femora with less protuberant tubercles on external pagina 2
2. Vertex sufficeant elevated above the pronotum surface; shoulder less than the above mentioned; third pulvilli a little longer the second .... *dorsifera* Walker
- Vertex medium elevated, Pronotum usually with a more or less distinct transvers black bands or two large spot just behind level of hind coxa .... *tarsalis* karby

Family–Tetrigidae

Subfamily–Scelimeninae

***Scelimena harpago* Serville, 1839 (Plate 40, Figures 342, 343)**

Thick brown, dorsoventraly flatened; eyes round; antennae black, thin; pronotum long, dorsal side rough, on ventrolateral side of pronotum present a pointed spine, nearly straight; tegmina reduced; femur laterally flattened; dorsal and lateral carina well marked; on ventral side of femora four pointed spines present; lamellae of hind tibia very wide, modified for swimming, a thin web present around the tibia and up to first and second tarsal segments; wings thin. In female, head 4 mm; pronotum 22 mm;

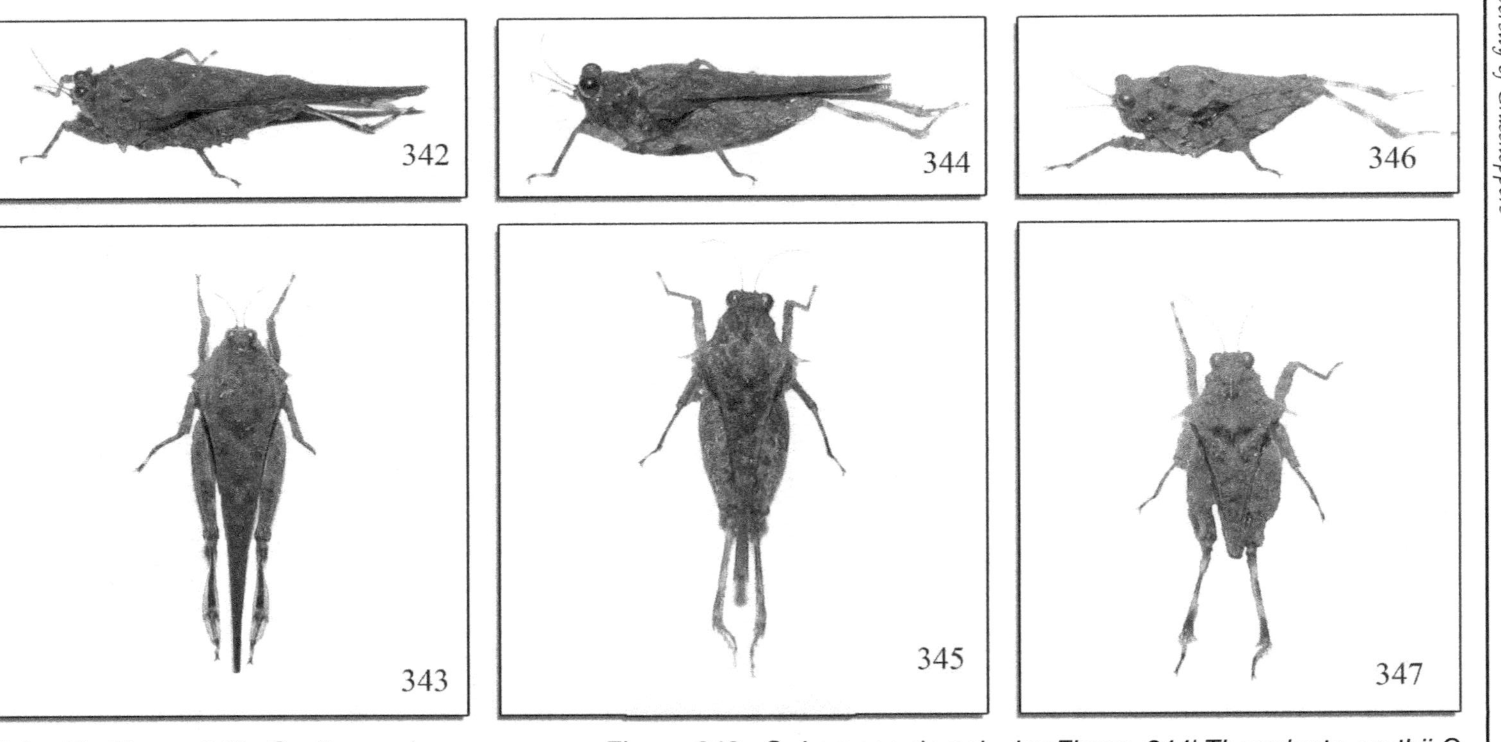

Plate 40: Figure 342: *Scelimena harpago.* serv.; Figure 343: *S. harpago* dorsal *viewFigure 344'.Thoradanta pruthii* G Figure 345: *T. pruthii* dorsal view; Figure 346: *Acanthalobus cuneatus* Hanc; Figure 347: *A. cuneatus*

prothorax 3 mm long; meso and meta thorax 5 mm long; abdomen 7 mm long; wing span 30 mm. In male, antennae 14 mm long; head 2.5 mm long; pronotum 18 mm long; prothorax 2 mm; meso and meta thorax 3 mm; abdomen 5 mm long; wing span 26 mm. It is semi-aquatic species found near the water bodies; feed on the aquatic vegetation algae and mosses.

Family–Tetrigidae

Subfamily–Scelimeninae

***Thoradonta purthii* Gunther, 1938 (Plate 40, Figures 344, 345)**

Brown; antennae, thin black hair like; head not exerted; pronotum rough, extend up to the tip of wing; lateral margin of pronotum behind the shoulders; in between front and middle legs small pointed spine is present. Posterior femora elongated and laterally compressed tibia faint brown, with short spine, two pale bands present on tibia. In male, antennae 2.5 mm; head 1.4 mm; thorax 3.5 mm; abdomen 3 mm; pronotum 9 mm long; body length 9.5 mm long including pronotum. In female, antennae 3 mm; head 2 mm; prothorax 4 mm long; pronotum 10 mm; abdomen 3.5 mm long.

Family–Tetrigidae

Subfamily–Scelimeninae

***Acantholobus cuneatus* Hancock, 1904 (Plate 40, Figures 346, 347)**

Colour brown; antennae hair like; eyes small, round; vretex broder than the eyes with slight median carina; frontal carina not prominent, pronotum short as compare to *Thoradonta purthi*; it is more rugose and more cuneate posteriorly, extend up to tip of femora, dorsal side is rough. A pointed spine is present in between fore and mid legs, hind femora rough, laterally compressed, two pale rings present on tibia, also row of small spines present on tibia; head 1.5 mm long; scape 0.7 mm long and 0.6 mm wide, pedicel 0.4 mm long and 0.4 mm wide; pronotum 8 mm long; body length including pronotum 8.5 mm long. This species feed on water melon, cucumber, algae, etc.

Flagellar formula – 1L/W =5, 4L/W =2.5, 7L/W = 7, A = 4.83

Family - Tetrigidae

Subfamily–Metrodorinae

***Systolederus cinereus* Brunner, 1893 (Plate 41, Figures 348,349)**

Very small, brown coloured species; head exerted above the pronotal surface; antennae thin; eyes round; pronotum, slightly convex, rough; legs brown; hind femora laterally fattened, brown coloured; hind tibia pale. Frist two pulvilli of the hind tarsi are equal length, the third longer; head 1.6 mm long; antennae 5.9 mm, scape 0.6 mm long and 0.2 mm wide, pedicel 0.3 mm long and 0.2 mm wide, flagella 5 mm long; pronotum 4 mm long; abdomen 2.2 mm long; hind femora 3 mm long; hind tibia 2.8 mm long. A pointed short spine present on ventrolateral side of pronotum in between fore and mid leg coxa. This species feed on algae and mosses.

Flagellar formula - 1L/W = 1.5, 4L/W =2, 7L/W = 6, A = 3.1

Family - Tetrigidae

Subfamily–Tetriginae

***Euparatettix personatus* (Bolivar, 1887) (Plate 41, (Figures 350, 351)**

Medium sized dark brown or grey coloured; vertex narrower than the eyes and elevated forward, paired ocelli placed in between middle of eyes. antennae some what long, located at the lower border of eyes; pronotum narrower, pointed at the extrmate, some time bordered with yellow line, median carin well marked; Wings long, extending beyond the pronotal apex; front femora slender hind femora elongated; tibia at apex light brown; first segment of posterior tarsus longer than the third segment.

Family - Tetrigidae

Subfamily–Tetriginae

***Hedotettix gracilis,* De Haan., 1842 (Plate 41, Figures 352, 353)**

Medium sized; head black; antennae small, hair like; eyes prominent, round; vertex wider than an eye, pronotum pale coloured, beyond the hind femora, median carina a little raised, wings longer than the pronotum; legs grey coloured, a black mark present dorsal side of tegmina just above the hind coxa. In female, antennae 3 mm long; head 2 mm long; pronotum 12 mm long; wings 2 mm longer than the pronotum; hind femora 5.5mm long; hind tibia 5mm long. In male,head 1.8 mm long; antennae 20.2 mm,

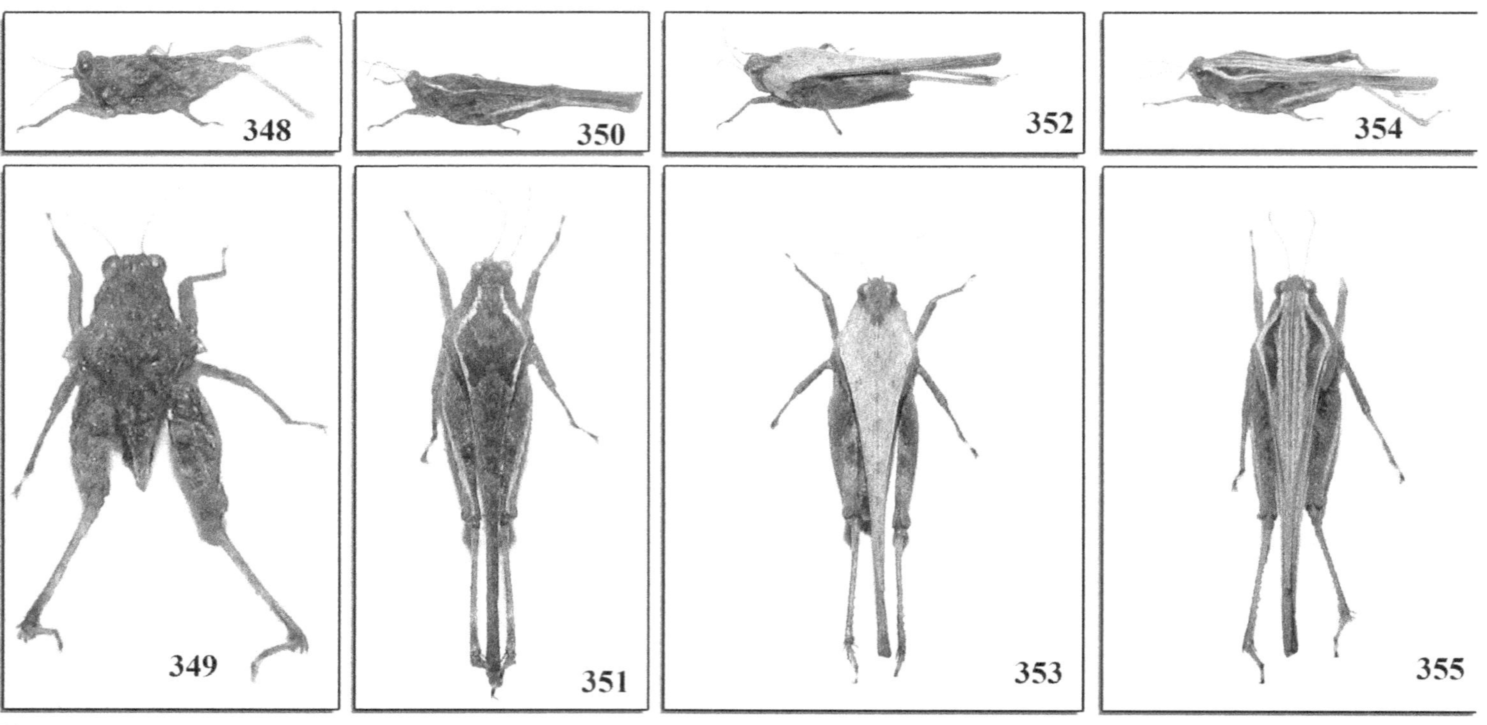

Plate 41: Figure 348: *Systolederus cinereus* Brunner; Figure 349: *S. cinereus* dorsal view; Figure 350: *Euparatettix personatus,* (Bolivar); Figure 351: *E. personatus* dorsal view; Figure 352: *Hedotettix gracilis* (De.Hann); Figure 353: *H. gracilis* dorsal view; Figure 354: *Hedotettix linnefera* Hanc.; Figure 355: *H. linnefera*

scape 2.2 mm long and 2 mm wide, pedicel 1mm long and 1.5 mm wide, flagella 17 mm long; pronotum 9 mm long; wings 2.5 mm longer than the pronotum; meso and meta thorax 2.5 mm long; abdomen 3 mm long; hind femora 5 mm long; tibia 4 mm. This species feed on wetland vegetation, newly emerged grasses algae and fungi.

Flagellar formula = 1L/W = 2.2, 4L/W =0.9, 7L/W = 0.9, A = 1.8

Family – Tetrigidae

Subfamily–Tetriginae

***Hedotettix lineifera* Walker, 1904 (Plate 41, Figures 353, 355)**

Grey brown; eyes round, broad; vertex slightly upward of the eyes; pronotum slightly raised, dorsal surface of pronotum brown, whitish longitudinal lines present on dorsal side, lateral carina yellowish; legs dark brown, longitudinal yellow band present on hind femora; tibia pale with small 8-9 pointed spines. In female, head 2.5 mm long; antennae 20.2 mm, scape 2.2 mm long and 2 mm wide, pedicel 1 mm long and 1.5 mm wide, flagella 0.9 mm long; pronotum 12.5 mm long; wings 2 mm longer than the pronotum; abdomen 3.5 mm long; hind femora 5 mm long; hind tibia 4.5 mm long.

Flagellar formula – 1L/W = 5, 4L/W =2.5, 7L/W = 7, A = 14.5

Family – Tetrigidae

Subfamily–Tetriginae

***Tetrix bipunctata* (Linnaeus, 1758) (Plate 42, Figures 356, 357)**

Blackish grey colour; antennae pale, black at the tip; eyes round; vertex larger than the eyes; pronotum obtuse angular anteriorly, cuneate posteriorly extend up to the apex of abdomen; wings up to the tip of pronotum; median carina sharp, lateral carina of pronotum also well marked, a black dot present on dorsolateral side at anterior tip of pronotum, also oblique groove present on dorsal side of pronotum just above the hind coxa; legs blackish grey; hind femora flat, whitish, sculpturing present on dorsal side of hind femora. Tibia pale coloured, 9-10 small pointed spines present on lateral side of hind tibia. In female, antennae 3.5 mm long; head 2.5 mm long; pronotum 10 mm long; meso and meta thorax 3 mm long; hind femora 6 mm long; hind tibia 5 mm long. This species feed on wetland weeds, algae, fungi, etc.

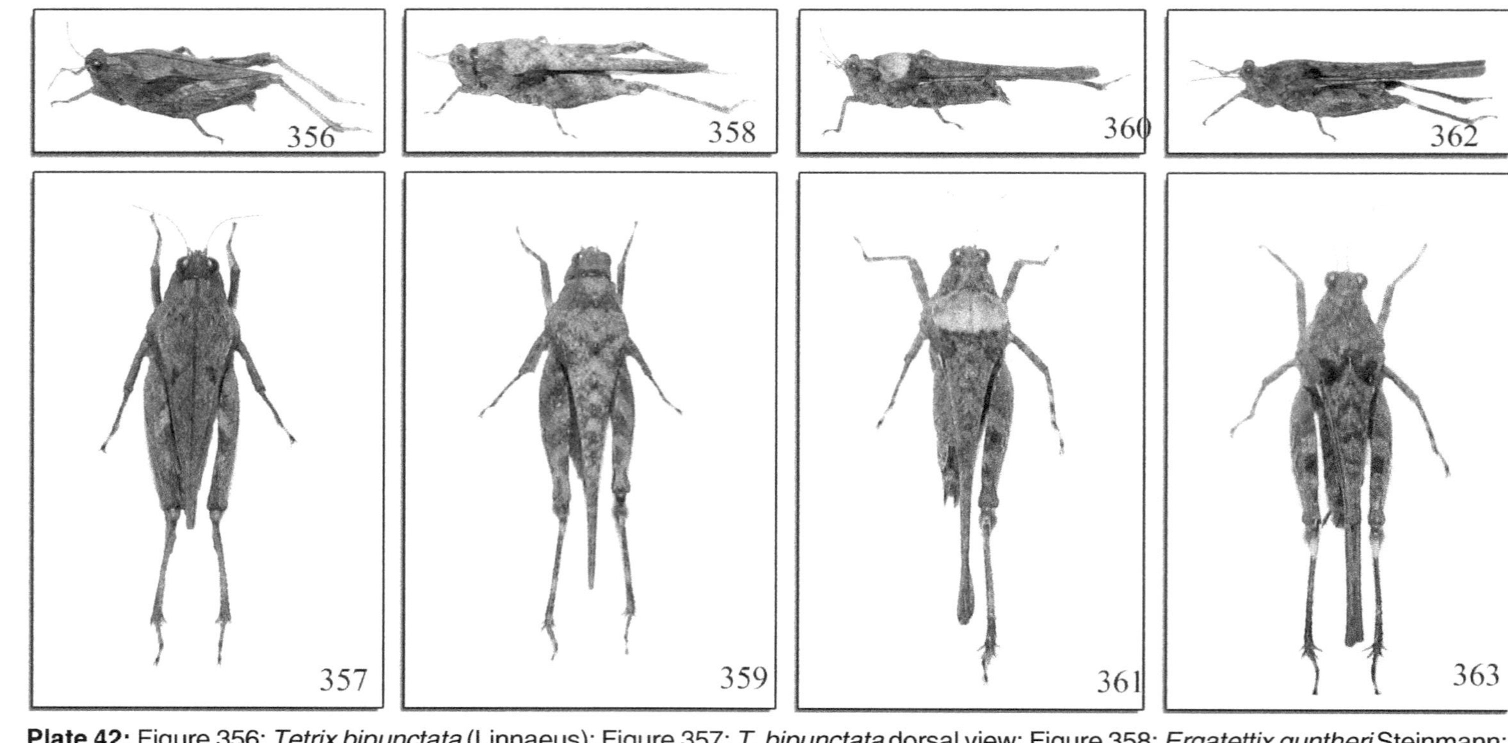

Plate 42: Figure 356: *Tetrix bipunctata* (Linnaeus); Figure 357: *T. bipunctata* dorsal view; Figure 358: *Ergatettix guntheri* Steinmann; Figure 359: *E. guntheri* dorsal view; Figure 360: *Ergatettix dorsifera* (Walker); Figure 361: *E. dorsifer* dorsal view; Figure 362: *Ergatettix tarsalis* Kirbay; Figure 363: *E. tarsalis* dorsal view

Family – Tetrigidae

Subfamily–Tetriginae

Ergatettix dorsifer **(Walker, 1871) (Plate 42, Figures 358, 359)**

Brown colour; antennae hair like; eyes round, at the tip of pronotum whitish band present, pronotum beyond the hind femora; tibia faint brownish; hind femora rough. In female, head 1.2 mm long; scape 0.6 mm long and 0.5 mm wide, pedicel 0.2 mm long and 3 mm wide, flagella 17 mm long; pronotum 11 mm; thorax 3.5 mm; abdomen 4 mm; body length including pronotum 12 mm. Feed on wetland weeds, cucumber, algae, mosses, etc.

Flagellar formula - 1L/W = 2, 4L/W =4, 7L/W =3.66, A = 3.2

Family – Tetrigidae

Subfamily–Tetriginae

Ergatettix tarsalis **Kirbay, 1914 (Plate 42, Figures 362, 363)**

Brown colour; antennae black towards the tip and pale towards the base; eyes round; antennae inserted distinctly below the inferior margin of eyes; vertex medium elevated; two black spots present on pronotum just behind the hind coxa. On hind femora two black dots present; hind tibia black with pointed spines, white ring present at the base. In female, antennae 4 mm long; pronotum 10 mm long; thorax 30 mm long; abdomen 2.5 mm long. Feed on wetland weeds, algae, mosses, etc.

Family – Tetrigidae

Subfamily–Tetriginae

Ergatettix guntheri **Steinmenn, 1970 (Plate 42, Figures 358, 359)**

Grey coloured; antennae thin hair like; eyes small, round; head greyish; pronotum long, beyond the hind femora; median carina of pronotum distinctly undulate, median femora sufficiently flattened, hind femora dark brown, strongly tuberculate; tibia pale; wings longer than the tegmina. In female, head 2 mm long; antennae 8.8 mm, scape 1mm long and 0.6 mm wide, pedicel 0.4 mm long and 0.3 mm wide, flagella 7.4 mm long; pronotum 11 mm; meso and meta thorax 3 mm long; hind femora 5 mm long; tibia 4 mm long; abdomen 3.5 mm long. In male, antennae 3 mm long; head 1.8 mm long; pronotum 10 mm long; abdomen 3 mm long; hind femora 4.6 mm long; hind tibia 4 mm long.

Flagellar formula – 1L/W = 1.66, 4L/W =5, 7L/W = 9, A = 5.22.

Key to the Subfamilies of Family Tettiigoniiidae

1. Head conical; tegmina narrow towards the tip; shorter than the wing. .... Conocephalinae
- Head conical tegmina narrow towards the tip, equal or slightly longer than the wing; body unicoloured .... Copiphorinae
2. On fore, mid and hind leg tibia present a spine, stridulatory organ well developed in male; on dorsal side of pronotum present brownish marking; basal tip of pronotum not round .... Listrosclidinae
- On fore and mid leg tibia spine absent, hind femora stout at base, basal of pronotum round .... Mecopodinae
3. Tegmina shorter than the wing stridulatory organ well developed; Ovopositor short, curved .... Phaneropterinae
- Body stout, some what flattened Stridulatory mechanism not well developed body appandesess irregular shape .... Pseudophyllinae
4. Hind tibia narrow, ventrolateral side of hind femora present a row of dorsolateral spines; stridulatory part of tegmina semitransparent .... Oxylakina

Key to the Genera of Subfamily Phaneropterinae

1. Pronotum dorsally smooth 2
- Pronotum dorsally not smooth 1
2. Body delicate; small black spot all over the body; tegmina with thick venation and narrow towards the tip .... *Lentena* Walker
- Body medium, tegmina not with thick venation and not narrow towards tip .... *Ductia* Thunberg
3. Lateral angle of pronotum rounded; tegmina long, narrow the tip, radius branching little before middle of tegmina .... *Tapiena* Ing. and Shis.

Family–Tettiigoniidae

Subfamily–Conocephalinae

Conocephalus maculatus **Leguillov (Plate 43, Figures 364 to 371)**

Greenish brown; head conical; eyes round and brown, a black strip run over the dorsal side of head and pronotum (Figures 367,368); lateral side of pronotum green; tegmina (Figure 369) brown, small black patches present on lateral side of tegmina. In female, a saw like ovipositor present; femora (Figure 370) broad at the base and narrow towards the tip; a row of lateral spines present on lateral side of tibia (Figure 371). In female, antennae 50 mm long; head 5 mm long; thorax 5 mm long; abdomen 15 mm long; tegmina 22 mm long; wing span 52 mm long. This species feed on aphids, mealy bugs and other small insects.

Family–Tettiigoniidae

Subfamily–Copiphorinae

Euconocephalus incertus **(Walkar, 1869) (Plate 44, Figures 372 to 378)**

Green or pale colour; head conical; eyes small, round mustard like; antennae long, pale; lateral carina of pronotum (Figures 374, 375) remarkable; tegmina (Figure 376) semitransparent, equal to the hind femora; wings transparent; hind femora (Figure 377) narrow, four small toothed spine shown in ventral side of hind femora towards the tip. A row of spines present on the lateral and ventral side of tibia (Figure 378), one black spot present at just behind the joint of femora and tibia. In female, antennae 36 mm long; head 10 mm long; thorax 8 mm long; abdomen 50 mm (including ovipositor); body length 62 mm long; tegmina 30 mm long; wing span 63mm; in male, antennae 18 mm long; head 6mm long; thorax 6 mm long; abdomen 10 mm long; tegmina 25 mm long; wing span 53 mm.

Family–Tettiigoniidae

Subfamily – Listroscelidinae

Hexacentrus unicolar **Serville, 1831 (Plate 45, Figures 379, 386)**

Body green; a brown band present over the head and pronotum; eyes round green; mandible pointed; five to six long pointed spines present on lateral side of fore and mid leg tibia; hind femora (Figure 385) green, broad at the base and narrow towards the tip; hind tibia (Figure 386) long with a row of lateral small spines, also row of spines present on ventral side of

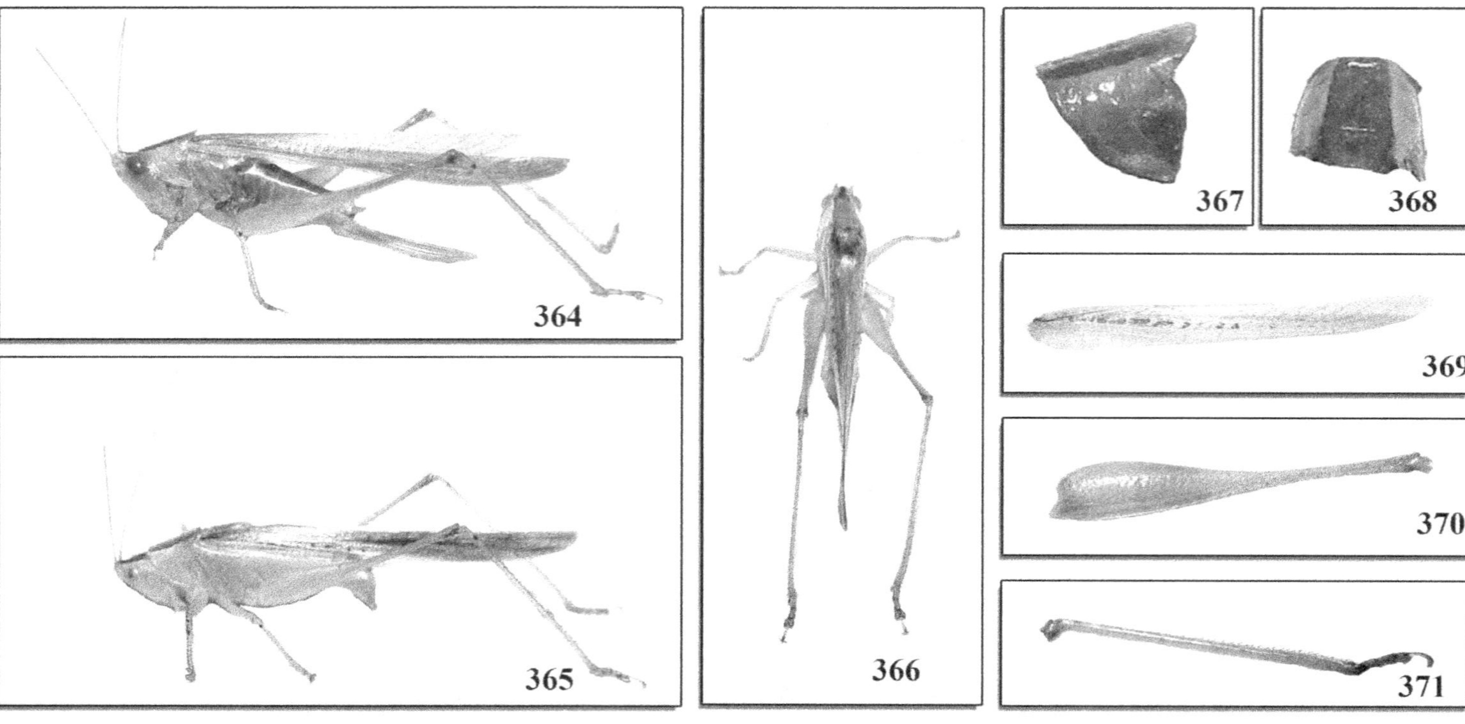

Plate 43: Figure 364: *Conocephalus maculatus* (Le Gull) Female; Figure 365: C. *maculatu* male; Figure 366: *C. maculatus* dorsal view; Figure 367: Lateral view of pronotum; Figure 368: Dorsal view of pronotum; Figure 369: Femur; Figure 370: Tegmina; Figure 371: Tibia

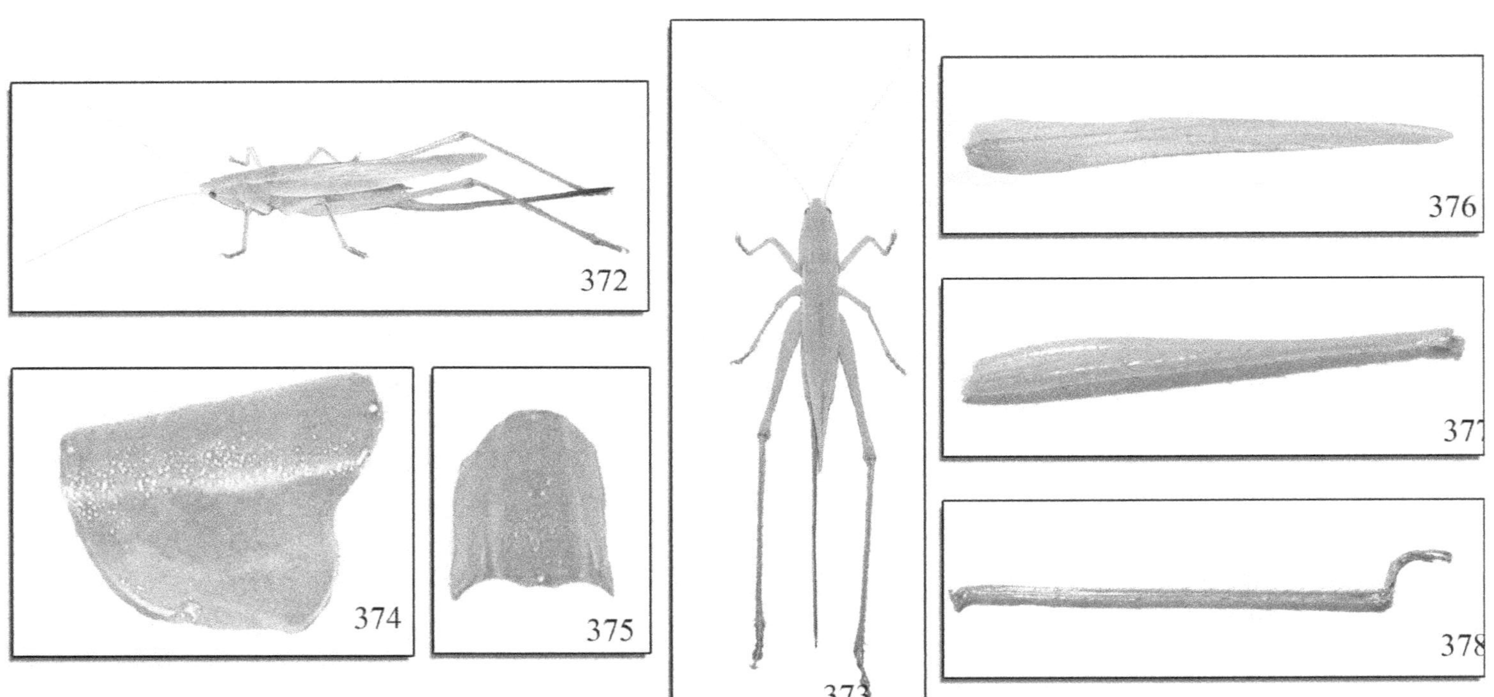

Plate 44: Figure 372: *Euconocephalus incertus* (Walker) Female; Figure 373: *E. incertus* dorsal view; Figure 374: Lateral view of pronotum; Figure 375: Dorsal view of pronotum; Figure 376: Tegmina; Figure 377: Femur; Figure 378: Tibia

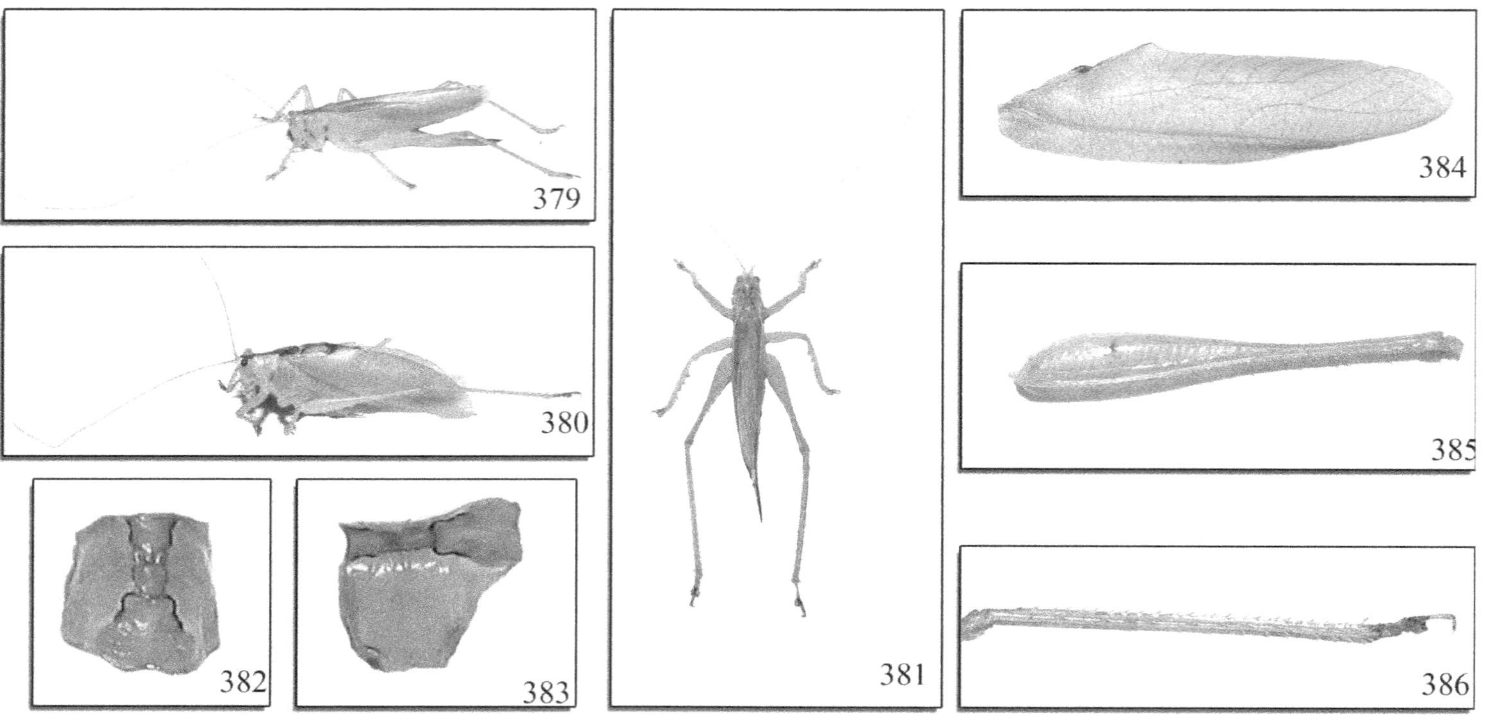

Plate 45: Figure 379: *Hexacentrus unicolor* Serville, Female; Figure 380: *H. unicolor* male; Figure 381: *H. unicolor* dorsal view; Figure 382: Pronotum dorsal view; Figure 383: Pronotum lateral view; Figure 384: Tegmina; Figure 385: Femur; Figure 386: Tibia

tibia; stridulatory organ present just behind posterior tip of pronotum. Male (Figure 380) of *H. unicolar* continuously produce monotonous sound at 7pm to 11 pm for attracting the female. It is very aggressive in nature. In male, antennae 52 mm long; head 7 mm long; thorax 8 mm long; abdomen 12 mm long; tegmina 38 mm long; wing span 76 mm long. In female, antennae 48 mm long; head 7 mm long; thorax 6 mm long; abdomen with ovipositor 30 mm long; tegmina 35 mm long; wing span 72 mm. It is truly predatory grasshopper, its hunting behaviour is very interesting, mostly found in long grasses and hunt the lepidopterous caterpillars and other small insects.

Family–Tettiigoniidae

Subfamily – Mecopodinae

***Mecopoda elongata* (Linnaeus, 1758) (Plate 46, Figures 387 to 393)**

It is large tettigoniid species; eyes semi oval, thick, brown; frontal ocelli yellow, present on frontal ridge; antennae long; pronotum (Figured 389, 390) brownish, lateral carina sharp, longitudinal black patch present on dorsolateral side of pronotum; stridulatory organ strong wings slightly longer than tegmina. The tegmina (Figure 391) narrow at the tip; wings pointed at the tip; hind femora (Figure 392) long, broad at the base, narrow towards the tip, ventrolateral side of femora present a row of pointed backwardly directed 5-6 spines; hind tibia (Figure 393) brown. A row of pointed black tipped spines present on lateral side of tibia, also a single row of pointed spines with basal black patch present on ventral side of tibia. In male, head 8 mm long; thorax 10 mm long; abdomen 20 mm; body length 34 mm long; tegmina 60 mm long; wing span 130 mm. The female more or less similar to the male only difference is that long ovipositor present at the tip of abdomen and white spot present on just above the midiastinal vein. This species highly nocturnal, male produce the monotonous sound for attracting the female at night only.

Family–Tettiigoniidae

Subfamily–Phnaeropterinae

***Lentena intermedia* Ingrich, 1990 (Plate 47, Figures 394 to 400)**

Eyes round, mustard like, brown; body green; very delicate; small black spot present over the body; antennae thin with black spot present; dorsal side of stridulatory organ is rough, brown coloured; tegmina (Figure 398)

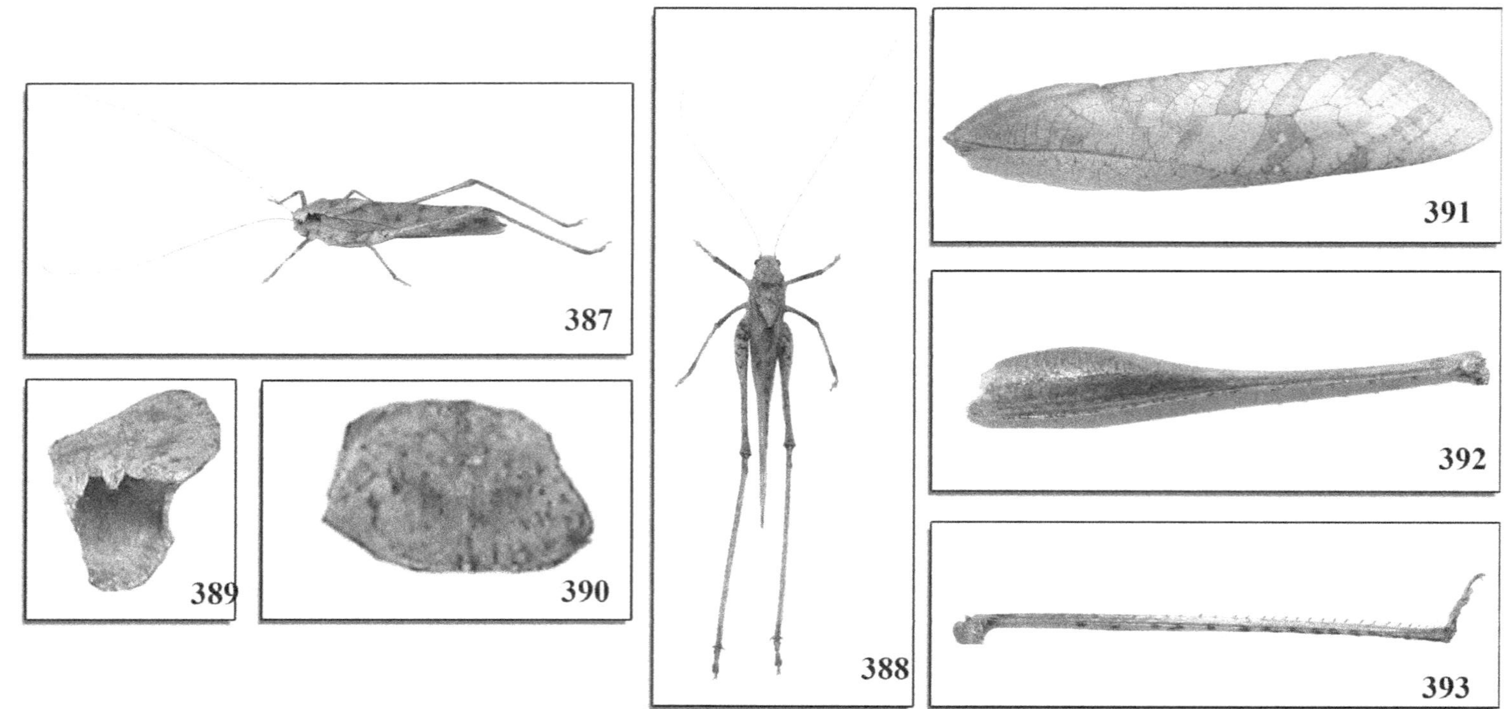

Plate 46: Figure 387: *Mecopoda elongata* (L.); Figure 388: *M. elongata* dorsal view; Figure 389: Pronotum lateral view; Figure 390: Pronotum dorsal view; Figure 391: Tegmina; Figure 392: Femur; Figure 393: Tibia

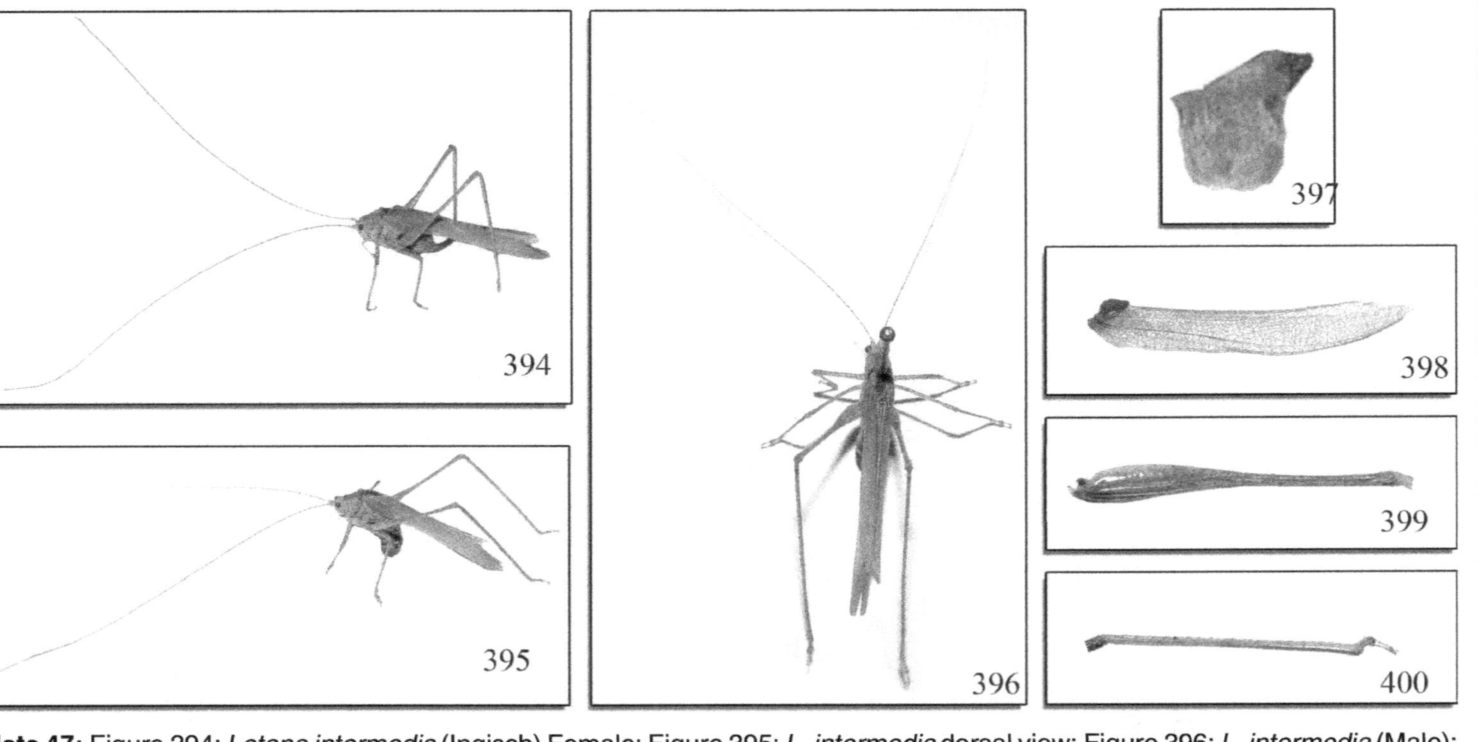

Plate 47: Figure 394: *Letana intermedia* (Ingisch) Female; Figure 395: *L. intermedia* dorsal view; Figure 396: *L. intermedia* (Male); Figure 397: Pronotum lateral view; Figure 398: Tegmina; Figure 399: Femur; Figure 400: Tibia

green shorter than the wings; hind femora (Figure 399) long, broad at the base; hind tibia (Figure 400) long, pointed spines present on the lateral side. In female, antennae 65 mm long; head 3.5 mm long; thorax 4 mm long; tegmina 30 mm long; wing span 74 mm; abdomen with ovipositor 14 mm long. In male, antennae 63 mm long; head 3 mm long; thorax 4 mm long; abdomen 8 mm long; tegmina 25 mm long; wing span 62 mm.

Family–Tettiigoniidae

Subfamily–Phnaeropterinae

***Ductia japonica* (Thunberg, 1878) (Plate 48, Figures 401 to 408)**

Body universaly green, except stridulatory region; tegmina (Figure 406) shorter than the wing; body elongated; eyes very small, round; antennae verymuch long, pedicel broad, median carina of pronotum brownish in male; legs greenish brown, a row of backwardly directed spines present on lateral side of tibia; tarsus three segmented, at the tip present claws and arolium, oviposition short and curved. In male head 5 mm long, pronotum 6 mm long; tegmina 3.8 mm long, hind femora 2.8 mm long, tibia 3 mm long and in female head 4.7 mm long, pronotum 5 mm long, tegmina 2.8 mm long, hind femora 2.5 mm long, hind tibia 2.7mm long.

Family–Tettiigoniidae

Subfamily–Phnaeropterinae

***Tapiena latifolia* Ingrisch and Shishodia, 2000 (Plate 49, Figures 409 to 414)**

Green; head small; eyes round, brown coloured; pronotum dorsaly flat, apex rounded, lateral angle of pronotum rounded; tegmina long narrow at the tip, radius branching little before middle of tegmina; tibia tympana open on external, anterior tibia with dorsal angle angular; without dorsal spines and with two to three ventroexternal and four ventro internal spines, dorsointernal spure absent. At ventrolateral side of hind femora present a reduced pointed backwardly directed six spines. On lateral side of hind tibia present a row of thick spines, also a row of spines present on ventrolateral side of hind tibia. In male antennae 70 mm long, pronotum 7.5 mm long and 4 mm wide; Tegmina 47 mm long, 13 mm wide, hind femora 27 mm long, hind tibia 32 mm long; hind tarsus 3.5 mm long.

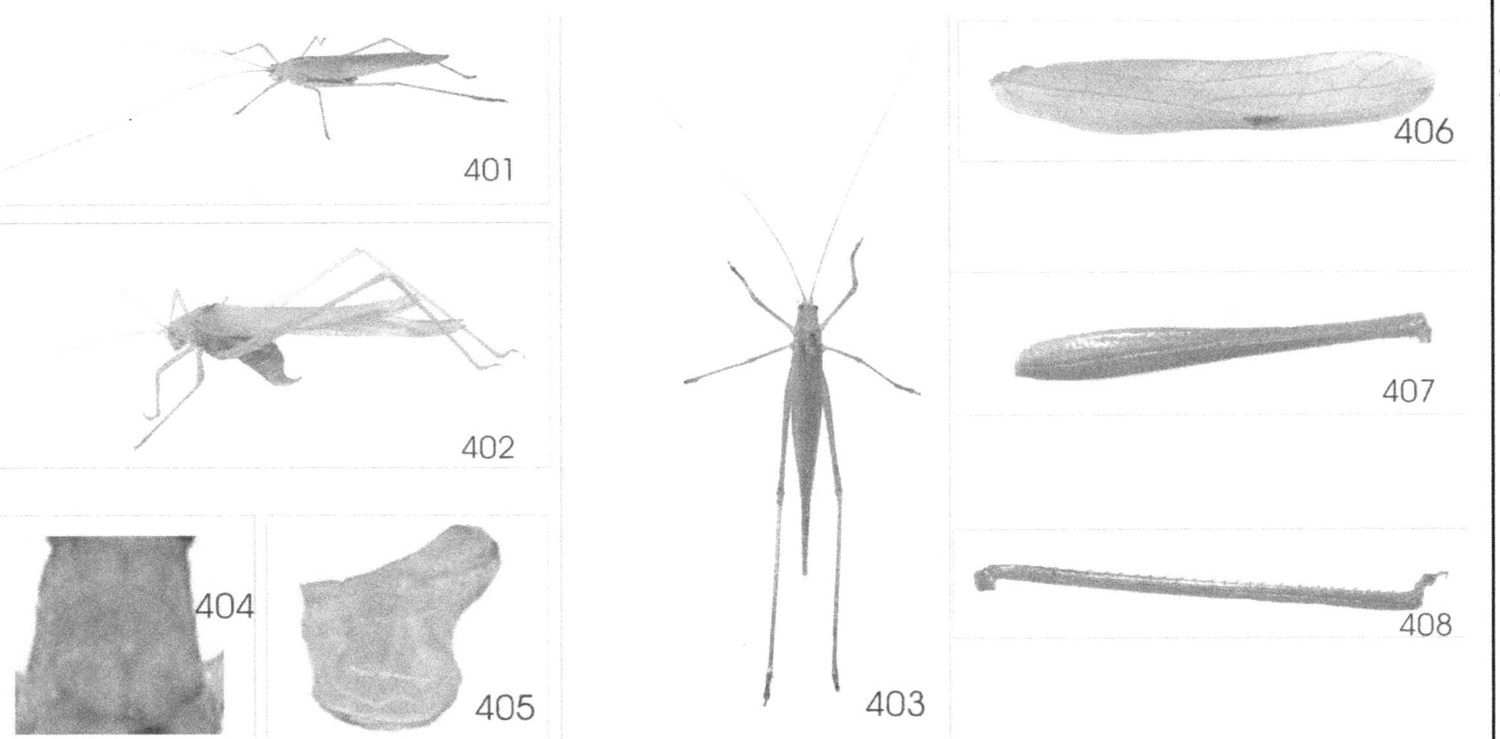

Plate 48: Figure 401: *Ductia japonica* (Thb.) male; Figure 402: *D. japonica* (Female); Figure 403: *D. japonica* dorsal view; Figure 404: Pronotum dorsal view; 405: Pronotum lateral view; Figure 406: Tegmina; Figure 407: Femur; Figure 408; Tibia

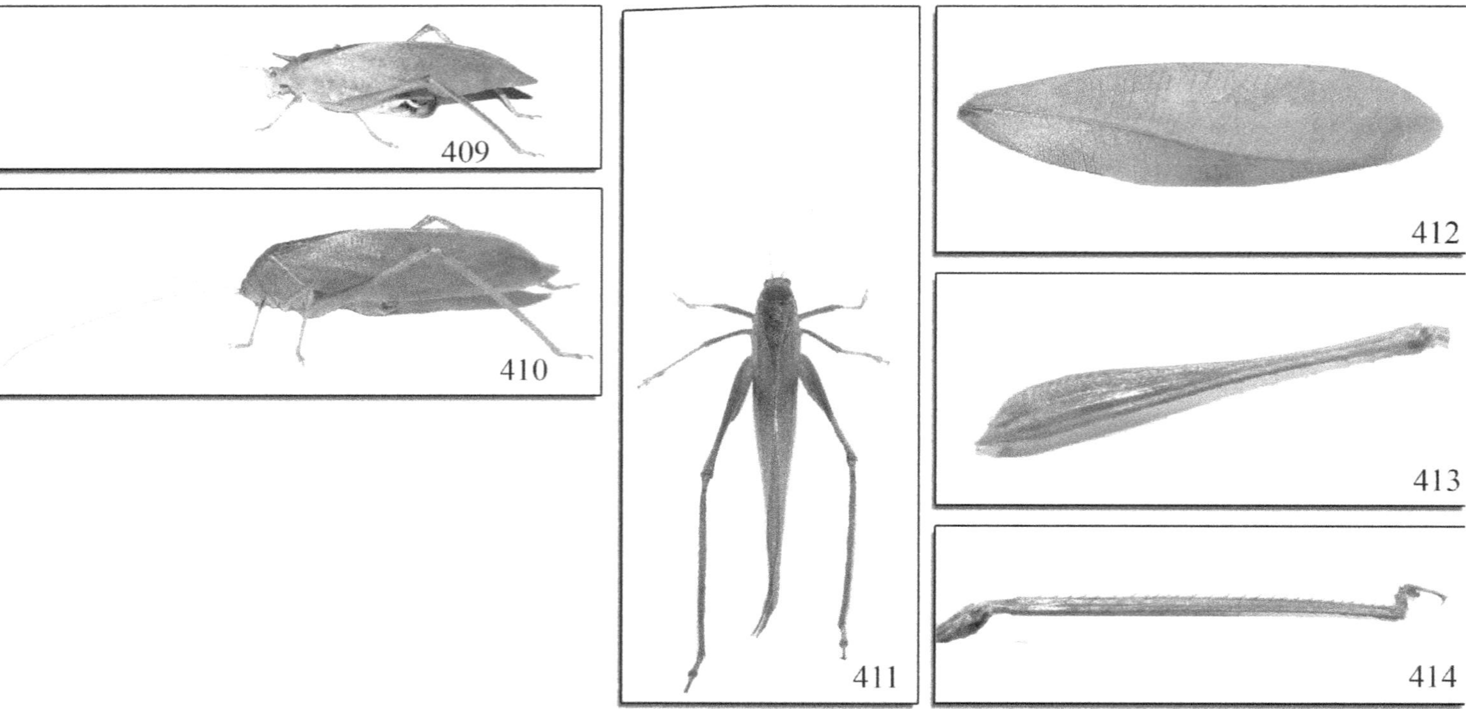

Plate 49: Figure 409: *Tapiena latifolia* (Ingrisch & Shishodia) female; Figure 410: *T. latifolia* (Male); Figure 411: *T. latifolia* dorsal view; Figure 412: Tegmina; Figure: 413: Femur; Figure 414: Tibia

Family–Tettiigoniidae

Subfamily–Psedophyllinae

Sathrophylla rugosa **(Linnaeus, 1758) (Plate 50, Figures 415 to 422)**

Grey coloured; dorsoventraly flattened; eyes upwardly directed, round; head semi round; antennae long, alternatively white and black band present on antenna; pronotum (Figures 419, 420) rough, frontal side of pronotum slightly spread over the head; lateral and median carina sickle like, shown only in posterior side of pronotum. In female, smokey whitish band shown on the tegmina (Figure 418) and femur (Figure 421) of hind legs. Legs flattened irregular shape. On fore legs femora present a sickle like web present at posteroventral side. On mid leg femora also present the sickle like web at frontoventral side and hind femora laterally flattened, ventral side like sickle; tibia (Figure 422) also irregular shape; wings longer than the tegmina. In female, brown black coloured saw like ovipositor present. In female, antennae 60 mm long; head 7 mm long; pronotum 8 mm long; tegmina 54 mm long; thorax 11 mm long; abdomen with ovipositor 35 mm long; body length 50 mm and wing span 114 mm. In male, antennae 50 mm long; head 7 mm long; pronotum 6 mm long; meso and meta thorax 20 mm long; tegmina 40 mm long; wing span 102 mm. It is nocturnal species sit on the stem of the plant and branches of plant, highly camouflagic, feed on Ber, Badam, Arjun and Ain like plant.

Family–Tettiigoniidae

Subfamily–Oxylakina

Kirkaldyus truncate **Karny, 1908 (Plate 51, Figures 423 to 429)**

Pale brown colour; eyes brown; semioval; antennae pale; head conical, longitudinal brown band present on lateral carina of head; pronotum(Figure 425, 426) smooth. A brown strip run dorsolateraly from just behind the eyes to some part of tegmina, stridulatory part of tegmina is semitransparent; hind tibia (Figure 429) narrow, ventrolateral side of hind femora (Figure 428) present a row of dorsolateral spine. On ventral side of tibia present a row of reduced spines. In male, head 9 mm long; antennae 45 mm long; pronotum 7 mm long; thorax 7 mm long; abdomen 12 mm long; tegmina 42 mm long; wing span 84 mm long.

Plate 50: Figure 415: *Satrophyllia rugosa* (Linnaeus), Male; Figure 416: *S. rugosa* Female; Figure 417: *S. rugosa* dorsal view; Figure 418: Tegmina; Figure 419: Pronotum dorsal view; Figure 420: Pronotum lateral view; Figure 421: Femur; Figure 422: Tibia

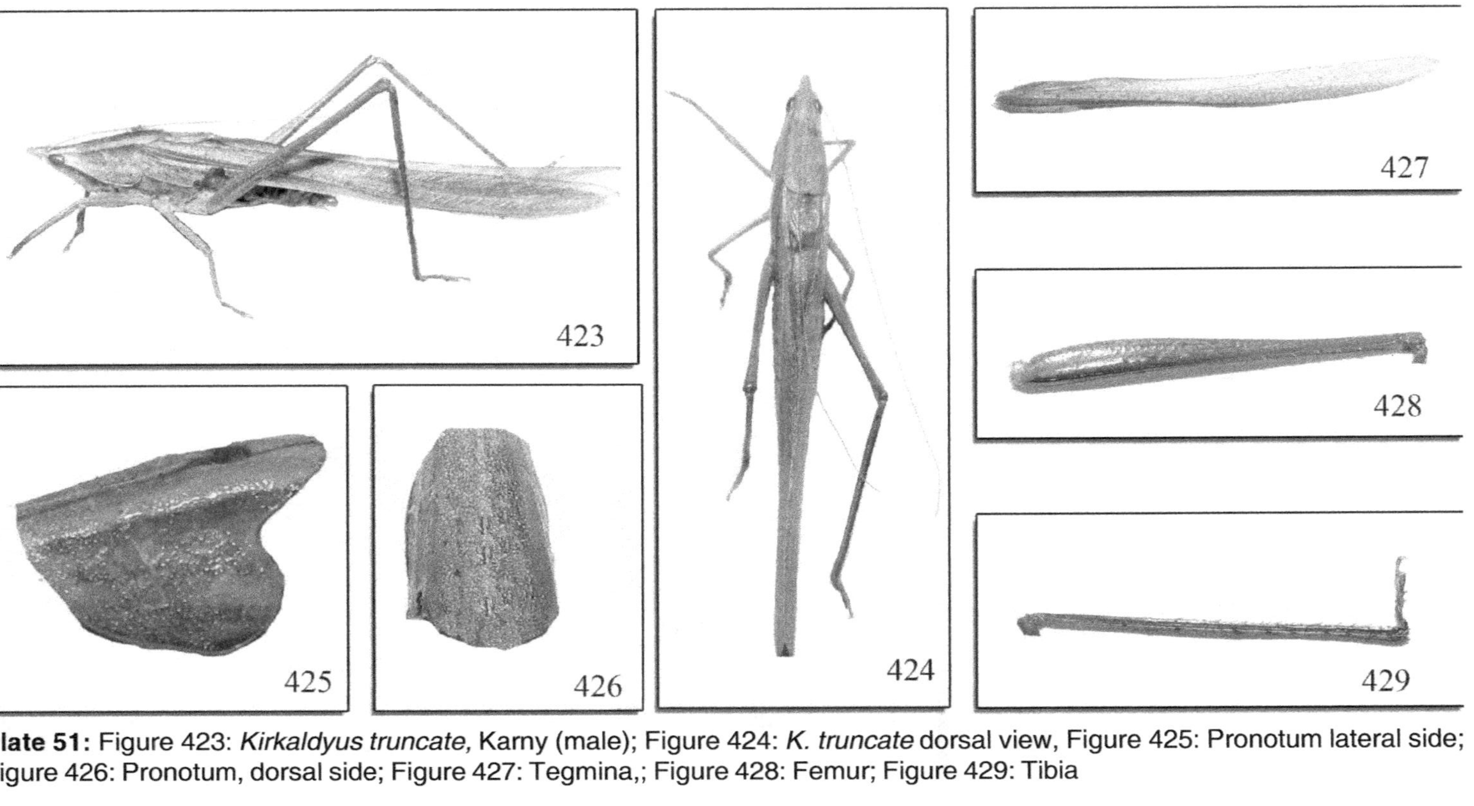

Plate 51: Figure 423: *Kirkaldyus truncate,* Karny (male); Figure 424: *K. truncate* dorsal view, Figure 425: Pronotum lateral side; Figure 426: Pronotum, dorsal side; Figure 427: Tegmina,; Figure 428: Femur; Figure 429: Tibia

Chapter 5
Seasonal Abundance, Distribution, Species Richness and Checklist of Grasshoppers

Introduction

Grasshoppers are injurious to several agricultural and forest crops. Hence, they should be studied with respect to ecology, seasonal abundance, distribution etc. The above aspects can play an important role in designing pests control programmes of grasshoppers. Ecological pest management has been attracted the attention of farmers and scientists, since it is ecofriendly. Thus geographical distribution seasonal abundance, habitat, natural enemies, climatic factors, etc. are important part of ecological pest management.

In population dynamics sampling of insect population is difficult and important task which estimates the number of species present in a target area. Seasonal abundance of a species suggests the time of occurrence and probable number of the species available in a specific region. Further, it helps in designing, appropriate control measures of target species and understanding useful role of insect species.

Calculating the species richness in a particular area is necessary for comparing the data between two different geographical areas. Biodiversity is the species richness in an ecosystem (Ehrlich and Ehrlich 1981). According to Jenkins (1992) the species richness data may provide relative little ecological significant information in practice, such are the most easily derived. Species richness is perhaps the most useful indices for comparison of diversity on a large geographical scale. Measures of biodiversity from particular areas, habitats or ecosystem are often largely reduced to a straight forward measure of species richness. Species richness should be measured by cataloging total species occurring in the selected area. According to Krishnamurty (2004) studying the species richness is practically unrealistic science even with a very small sites complete cataloging of all species available is not possible. Hence, Species richness is measured on samples carefully chosen in a particular area.

To study the species richness, two kinds of geographical areas, *viz.* Plain region and Ghats region were selected in the present work. Specially two tahsils were selected from Kolhapur district. Karveer (Figure 430) as a plain region and Radhanagry (Figure 431) as Ghats region

Karveer is the central region of Kolhapur district with brownish black coloured soil at Eastern region and Red latretic soil found at Western region of tahsil, it is plain and fertile area.

Agricultural crops such as rice, sugarcane, wheat are the major crops grown in the area. Maximum temperature ranges from 36°C to 40°C in April and minimum 20°C to 15°C in January. The Western part of Karveer includes the medium rainfall Zone (1450-2000 mm rain fall) received from the South West mansoon, between June to September. Maximum rain is received in months of July and August. The region contain dry mixed forests. Estern part of Karveer includes the low rain fall Zone (775- 900 mm) an dry deciduous forests. Radhanagari tahsil comes under Sahyadri ranges. This humid, wet, and with high rain fall and cool region with red soil covered by thick forest and located 1000 meters msl and recives about 6000 mm rainfall from the South West monsoon between June and September. Temperature ranges 35°C to 37°C during summer and 15°C to 10.2°C during winter. The tahsil contain mixed semi evergreen and moist deciduous forests showing very good floral and faunal diversity in the region. Important animals found in the region refer to Tiger, Panther, Wild Cat, Hyena, Jackal, Indian gray mongoose, Rubby mongoose, Wild Dog, Samber, Bear, Barking dear, Gaur, Slot bear, Giant squirrel, several birds, amphibians and reptiles with rich fauna and flora.

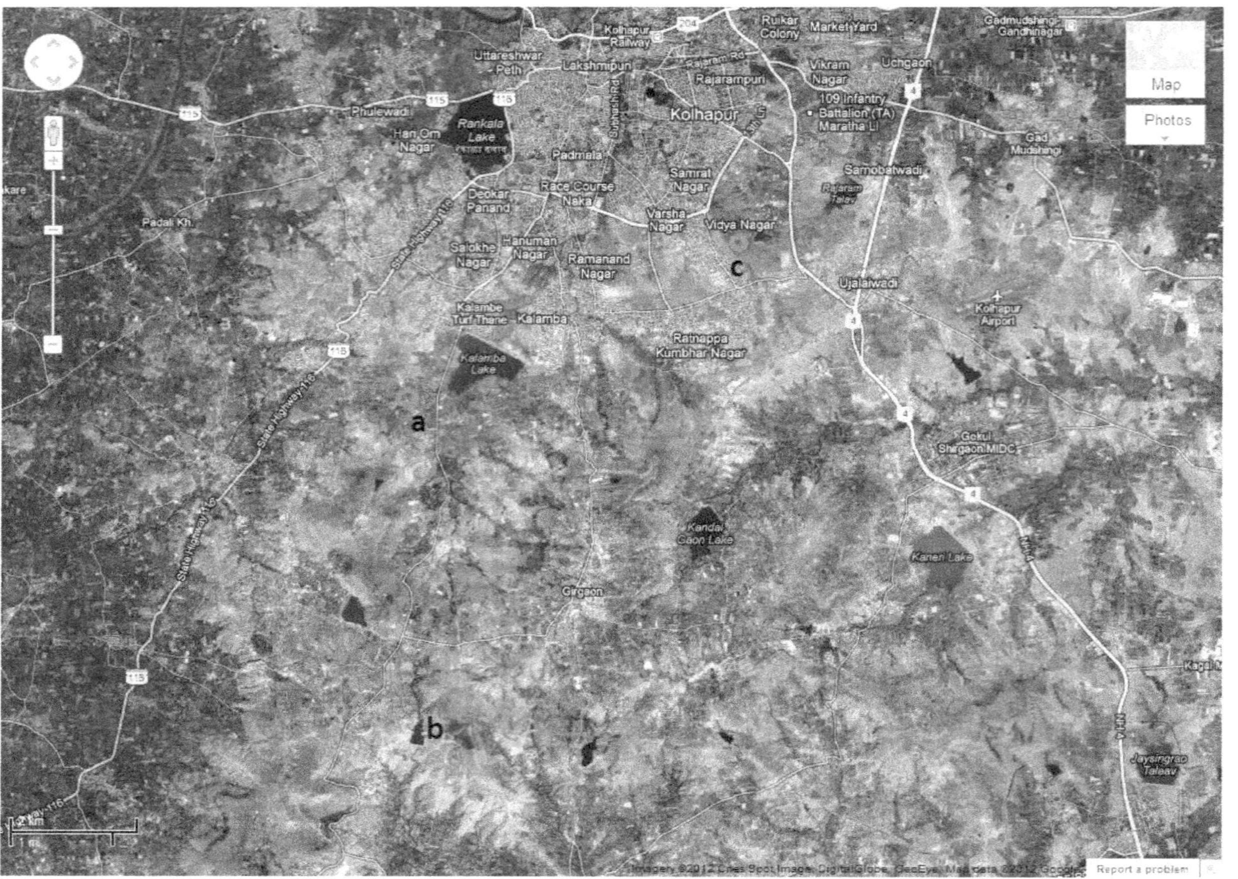

Figure 430: Satellite View of Study Area (Karveer) ◯ Study spot.

a: Agriculture; b: Grassland; c: Semiforest

Figure 431: Satellite View of Study Area (Radhanagari) ● Study spot.

a: Agriculture; b: Semiforest; c: Pleatue.

Perusal of literature indicates that very little attention is paid on the ecology, seasonal abundance and distribution of grasshoppers in India including Western Ghats. Ananthaselvi *et al.* (2006), Aswanthanarayan *et al.* (1981) Beason (1941), Bhowmik (1984), Chandra (2003, 2007), Chandra and Gupta (2009), Gardener (2005), Julka *et al.* (1982), Kandibane *et al.* (2004), Mayya (2003), Senthilkumar (2010), Shishodia (1991, 1995, 1999, 2000), Tondan (1995), Thakur (2011) etc have attempted the ecological studies of grasshoppers.

MATERIALS AND METHODS

The seasonal abundance of grasshoppers was studied in different places of Kolhapur district namely Shahuwadi, Gaganbawda, Radhanagari, Ajra, Gadhinglaj, Chandagad, Panhala, Karveer, Bhudhargad, Kagal and Hatkalangale. From Sangali district Miraj, Battish Shirala, Islampur, Tasgoan, Jath, Chandoli, Sagareshwar and Vita were selected while, from Satara district Koyna, sajjangad, Mahabaleshwar, Karad, Wai, Lonand and Pachgani were selected for the study of grasshopper seasonal abundance. Grasshoppers were collected with the help of insect net (Figure 3) for one man one hour search method during the years 2009-2012. The collected material was identified by consulting appropriate, literature and preserved in Department of Zoology, Shivaji University, Kolhapur.

Distribution records of grasshoppers have been made by visiting, collecting and identifying the species from study spots of Western Maharashtra (Figures 16, 17) and consulting previous distributional record of the region.

To Study the species richness from Ghats (Radhanagari) and Plain region (Karveer) following methods were followed. The species of family Acrididae and Pyrgomorphidae were recorded in 15x15 m quadrate from six different spots from Ghats and Plain region at 15 days interval, in the morning and evening hour during the years 2009-2011. Grasshopper richness was estimated by collecting common species by sweep net for about 30 minute in 15x15 meter area as described by Velez (2008). Collected species of grasshoppers are identified and counted. After completion of data collection, the arithmetic mean was calculated from the data of three different ecosystems *viz*. Agricultural, grassland and semiforest. The same data was used for graphical presentation of the species density. Later diversity was measured by using Spimpson index.

Spimpson Index:

$$D = \sum_{i=1}^{s} \frac{n_i(n_i - 1)}{N(N-1)}$$

where,

n is the total number of organism of a particular species and N is the total number of organism of all species.

Results

The abundance of species found in Plain region is recorded in Table 1. In all, 46 species of grasshoppers were reported belonging to four families, 19 subfamilies and 42 genera from Plain region of Western Maharashtra. Out of which 42 species were common and 4 species were rare. The family Acrididae contains four rare species *viz., G. semipictus, P. bilineatus, T. monticollis* and *C. robustus* and family Tetrigidae was reported by only one species namely *S. cinereus* During the study 26 species of family Acrididae were reported. Out of which 6 species *viz. A. exaltata, G. africanus africanus, T. annulata, S. prasiniferum prasiniferum, O. hyla hyla* (Serv.) and X. *humalis humails* were more adaptive because they survive all climatic conditions of the year. *T. indica* Bol., *D. desius, H. respondens, T. montocolis, S. splendens* (Th.) and *E. alacris alacris* were reported from late summer to late winter. *P. bilineatus, P. infumata, P. panteli, H. banian, A. flavescence, G. laticornis* and *G. semipictus* reported in pre-monsoon to mid winter period while, *A. thalassinus tamulus* (Fabr.), *O. hyla hyla* and *C. robustus* emerged in mid monsoon period and they live up to mid winter. *I. convergence, D. venustas, T. varicornis, P. sunccincta,* and *C. tatarica* specially adopted for dry climate and reported in summer season. Four species of family Pyrgomorphidae *viz. A. crenulata, C. oxypterus C. sphenarioidides* and *T. apludi* sp nov. were reported from July to Desember. *C. spenarioides* reported from August to December and *T. apludi* sp nov reported February to July. Ten species of family Tetrigidae reported from study area refer to *S. harpago, T. pruthii, A. cuneatus, E. personatus, E. dorsifer, E. tarsalis, E. guntheri, H. lineifera, H. gracilis* and *S. cinereus,* all are reported from

Table 1: Relative Abundance of Grasshoppers from Plain Region of Western Maharashtra

Sl.No.	*Name of Grasshoppers*	*Jan*	*Feb.*	*March*	*April*	*May*	*June*	*July*	*Aug*	*Sept*	*Oct.*	*Nov.*	*Dec.*
	Subfamily–Acridinae												
1.	*Truxalis indica* Bol.	–	–	–	–	+	+	+	+Δ	+	+	+	+0
2.	*Acrida exaltata* Walker	+	+	+	+0	+	+	+	+	+Δ	+	+	+
3.	*Phlaeoba panteli* Bol.	+0	–	–	–	–	+	+	+Δ	+	+	+	+
4.	*Phlaeoba infumata* Brunner.	+0	–	–	–	–	+	+	+	+	+Δ	+	+
	Subfamily–Gomphocerinae												
4.	*Gelostorrhinus laticornis* (Serv.)	–	–	–	–	–	–	+	+	+Δ	+	+	+0
5.	*Gelostorrhinus semipectus** Walk.	–	–	–	–	–	+	+	+	+	+	+	–
	Subfamily–Oedipodinae												
6.	*Dociostaurus desius* Walk	–	–	–	–	–	–	–	+	+Δ	+	+	+0
7.	*Stauradorus indica* sp. nov	–	–	–	–	–	–	+	+	+	+Δ	+	+0
8.	*Aiolopus thalassinus tamulus* (Fabr.)	–	–	–	–	–	–	+	+	+Δ	+	+	+0
9.	*Gostrimargus africanus africanus* Sauss	+	+	+	+0	+	+	+	+	+	+	+Δ	+
10.	*Heteropternis respondens* (Walk)	–	–	–	–	+	+	+	+Δ	+	+	+	+0
11.	*Peternoscirta sahyadricus* sp.nov.	–	+	+	+	+Δ	+	+0	–	–	–	–	–
12.	*Trilophidia annulata* (Th.)	+	+	+	+0	+	+	+	+Δ	+	+	+	+
	Subfamily Hemiacridinae												
13.	*Spathosternum prasiniferum prasiniferum* (Walk.)	+	+	+	+0	+	+	+	+	+Δ	+	+	+
14.	*Hieroglyphus banian*	–	–	–	–	–	–	+	+	+Δ	+	+	+0
15.	*Parahieroglyphus bilineatus** Kirbay	–	–	–	–	–	–	+	+	+	+	+	+

Contd...

Table 1–*Contd...*

Sl.No.	*Name of Grasshoppers*	*Jan*	*Feb.*	*March*	*April*	*May*	*June*	*July*	*Aug*	*Sept*	*Oct.*	*Nov.*	*Dec.*
	Subfamily–Oxynae												
16.	*Oxya hyla hyla* (Serv.)	+	+	+	+0	+	+	+	+	+Δ	+	+	+
17.	*Oxya japonica japonica* (Th.)	–	–	–	–	–	–	+	+	+Δ	+	+	+0
	Subfamily–Cyrtacanthacridinae												
18.	*Patanga succincta* (Johonsson)	+	+	+	+	+	+Δ	+	+	+0	–	–	–
19.	*Cyrtacanthacris tatarica* (Linn.)	+	+	+0	–	–	–	–	+	+	+	+Δ	+
	Subfamily–Romalinae												
20.	*Teratodes monticollis** Gray	–	–	–	–	+	+	+	+	+	+	+	+
	Subfamily–Catantopinae												
21.	*Catantops pinguise* Stal	–	–	–	–	+	+	+	+Δ	+	+	+	+0
22.	*Xenocatantops humalis humails* (Serv.)	+	+	+	+0	+	+	+	+	+Δ	+	+	+
23.	*Stenocatantops splendens* (Th.)	–	–	–	–	+	+	+	+Δ	+	+	+	+0
	Subfamily–Eprepocenemidinae												
24.	*Choroedocus robustus** (Serville)	–	–	–	–	–	–	+	+	+	+	+	+
25.	*Tylotropidius varicornis* (Walk)	+	+	+	+	+Δ	+	+0	–	–	–	–	–
26.	*Eyprepocenemis alacris alacris* (Serve.)	–	–	–	+	+	+Δ	+	+	+	+	+	+0
	Family Pyrgomorphidae												
	Subfamily–Orthacridinae												
27.	*Colemania sphenarioidides* Bolivar	–	–	–	–	–	–	–	+	+	+Δ	+	+0

Contd...

Table 1–*Contd...*

Sl.No.	*Name of Grasshoppers*	*Jan*	*Feb.*	*March*	*April*	*May*	*June*	*July*	*Aug*	*Sept*	*Oct.*	*Nov.*	*Dec.*
	Subfamily–Pyrgomorphinae												
28.	*Chrotogonus oxypterus* Blanchard	–	–	–	–	–	+	+	+Δ	+	+	+	+0
29.	*Atractomorpha crenulata* (Fabricius)	+0	–	–	–	–	+	+	+	+Δ	+	+	+
30.	*Tagasta apludi* sp. nov.	–	+	+	+	+Δ	+	+0	–	–	–	–	–
	Family Tetrigidae												
	Subfamily–Scelimeninae												
31.	*Scelimena harpago* Serv.	+	+0	–	–	–	–	–	+	+	+Δ	+	+
32.	*Thoradonta pruthii* Gunther	+	+0	–	–	–	–	–	+	+	+Δ	+	+
33	*Acanthalobus cuneatus* Hanc.	+	+0	–	–	–	–	–	+	+	+Δ	+	+
	Subfamily–Metrodorinae												
34.	*Systolederus cinereus** Brunn.	+	+	–	–	–	–	–	+	+	+	+	+
	Subfamily–Tetriginae												
35.	*Euparatettix personatus* Bol.	+	+0	–	–	–	–	–	+	+	+Δ	+	+
36.	*Hedotettix gracilis* De Haan.	+	+0	–	–	–	–	–	+	+	+Δ	+	+
37.	*Hedotettix lineifera* Walk.	+	+0	–	–	–	–	–	+	+	+Δ	+	+
38.	*Ergatettix dorsifer* (Walk)	+	+0	–	–	–	–	–	+	+	+Δ	+	+
39.	*Ergatettix tarsalis* Kirbay	+	+0	–	–	–	–	–	+	+	+Δ	+	+
40.	*Ergatettix guntheri* Stenmann	+	+0	–	–	–	–	–	+	+	+Δ	+	+

Contd...

Table 1–*Contd...*

Sl.No.	*Name of Grasshoppers*	*Jan*	*Feb.*	*March*	*April*	*May*	*June*	*July*	*Aug*	*Sept*	*Oct.*	*Nov.*	*Dec.*
	Family Tettiigonidae												
	Subfamily–Conocephalinae												
41.	*Conocephalus maculates*	+	+0	–	–	–	–	–	+	+	+Δ	+	+
	Subfamily–Copiphorinae												
42.	*Euconocephalus maculates* Walker	+	+0	–	–	–	–	–	+	+	+Δ	+	+
	Subfamily–Listroscelidinae												
43.	*Hexacentrus unicolar* Serville	–	–	–	–	–	–	+	+	+	+Δ	+	+0
	Subfamily–Mecopodinae												
44.	*Mecopoda elongata* (Linnaeus)	+0	–	–	–	–	–	+	+	+	+Δ	+	+
	Subfamily–Phnaeropterinae												
45.	*Lentena intermedia* Ingrich	–	–	–	–	–	+	+	+Δ	+	+	+	+0
46.	*Ducetia japonica* (Thunberg)	–	–	–	–	–	–	+	+Δ	+	+	+	+0

+: indicates that species presence; –: indicates that species absence, Δ: indicates that Maximum population, 0: indicates that minimum population, *: indicates that rare species.

Table 2: Relative Abundance of Grasshoppers from Ghats Region of Western Maharashtra

Sl.No.	Name of Grasshoppers	Jan	Feb.	March	April	May	June	July	Aug	Sept	Oct.	Nov.	Dec.
	Subfamily–Acridinae												
1.	*T. indica* Bol.	+	+0	–	–	+	+	+	+Δ	+	+	+	+
2.	*A. exaltata* Walker	–	–	–	+	+	+	+	+	+Δ	+	+	+0
3.	*P. panteli* Bol.	+0	–	–	–	–	+	+	+Δ	+	+	+	+
	Subfamily–Gomphocerinae												
4.	*G. laticornis* (Serv.)	–	–	–	–	–	–	+	+	+Δ	+	+	+0
5.	*Ischnarida convergence* * (Walk)	–	+	+	+	+	+	+	–	–	–	–	–
	Subfamily–Oedipodinae												
6.	*D. desius* Walk	–	–	–	–	–	–	+	+	+Δ	+	+	+0
7.	*Ditopternis venusta** (Walk)	+	+	+Δ	+	+0	–	–	–	–	+	+	+
8.	*S. indica* sp nov	–	–	–	–	–	–	+	+	+Δ	+	+	+0
9.	*A. thalassinus tamulus* (F.)	+	+	–	–	–	+	+	+	+Δ	+	+	+0
10.	*G. africanus africanus* Sauss	+	+	+	+	+Δ	+	+	+	+0	+	+	+
11.	*H. respondens* (Walk)	–	–	–	–	+	+	+	+Δ	+	+	+	+0
12.	*P. sahyadricus* sp.nov.	–	+	+	+	+Δ	+	+0	–	–	–	–	–
13.	*T. annulata* (Th.)	+	+	+	+	+	+	+	+Δ	+	+	+	+0
	Subfamily Hemiacridinae												
14.	*S. prasiniferum prasiniferum* (Walk.)	+	+	+	+0	+	+	+	+	+Δ	+	+	+
15.	*H. kohlapurensis* sp.nov.	–	–	–	–	–	–	+	+	+Δ	+	+	+0
16.	*P. bilineatus* Kirbay	–	–	–	–	–	–	+	+	+	+	+	+

Contd...

Table 2–*Contd...*

Sl.No.	Name of Grasshoppers	Jan	Feb.	March	April	May	June	July	Aug	Sept	Oct.	Nov.	Dec.
	Subfamily–Oxynae												
17.	*O. hyla hyla* (Serv.)	+	+	+	+0	+	+	+	+	+Δ	+	+	+
18.	*O. japonica japonica* (Th.)	–	–	–	–	–	–	+	+	+Δ	+	+	+0
	Subfamily–Cyrtacanthacridinae												
19.	*P. succincta* (Johonsson)	+	+	+	+	+	+Δ	+	+	+0	–	–	–
20.	*C. tatarica* (Linn.)	+	+	+	–	–	–	–	+	+	+	+	+
21.	*A. flavescence** (Fab.)	+	–	–	–	–	+	+	+	+	+	+	+
	Subfamily–Romalinae												
22.	*T. monticollis* Gray	–	–	–	–	+	+Δ	+	+	+	+0	+	+
	Subfamily–Catantopinae												
23.	*C. pinguise* Stal	–	–	–	–	+	+	+	+Δ	+	+	+	+
24.	*X. humalis humails* (Serv.)	+	+	+	+0	+	+	+	+	+Δ	+	+	+
25.	*S. splendens* (Th.)	–	–	–	–	+	+	+	+Δ	+	+	+	+0
	Subfamily–Eprepocenemidinae												
26.	*T. varicornis* (Walk)	–	+	+	+	+Δ	+	+0	–	–	–	–	
27.	*E. alacris alacris* (Serve.)	–	–	–	+	+	+Δ	+	+	+	+	+	+0
	Family Pyrgomorphidae												
	Subfamily–Orthacridinae												
28.	*C. sphenarioidides* Bolivar	–	–	–	–	–	–	–	+	+	+Δ	+	+0

Contd...

Table 2–*Contd...*

Sl.No.	*Name of Grasshoppers*	*Jan*	*Feb.*	*March*	*April*	*May*	*June*	*July*	*Aug*	*Sept*	*Oct.*	*Nov.*	*Dec.*
	Subfamily–Pygomorphinae												
29.	*C. oxypterus* Blanchard	+	+	–	–	–	–	+	+	+Δ	+	+	+
30.	*A. crenulata* (Fabricius)	+0	–	–	–	–	+	+	+	+Δ	+	+	+
31.	*T. apludi* sp. nov.	–	+	+	+	+Δ	+	+0	–	–	–	–	–
	Family Tetrigidae												
	Sub family–Scelimeninae												
32.	*S. harpago* Serv.	+	+0	–	–	–	–	–	+	+	+Δ	+	+
33.	*T. pruthii* Gunther	+	+0	–	–	–	–	–	+	+	+Δ	+	+
34.	*A. cuneatus* Hanc.	+	+0	–	–	–	–	–	+	+	+Δ	+	+
	Subfamily–Metrodorinae												
35.	*S. cinereus* Brunn.	+	+	–	–	–	–	–	+	+	+	+	+
	Subfamily–Tetriginae												
36.	*Tetrix bipunctata* *(Linnaeus)	+	+0	–	–	–	–	–	+	+	+Δ	+	+
37.	*E. personatus* Bol.	+	+0	–	–	–	–	–	+	+	+Δ	+	+
38.	*H. gracilis* De Haan.	+	+0	–	–	–	–	–	+	+	+Δ	+	+
39.	*H. lineifera* Walk.	+	+0	–	–	–	–	–	+	+	+Δ	+	+
40.	*E. dorsifer* (Walk)	+	+0	–	–	–	–	–	+	+	+Δ	+	+
41.	*E. tarsalis* Kirbay	+	+0	–	–	–	–	–	+	+	+Δ	+	+
42.	*E. guntheri* Stenmann	+	+0	–	–	–	–	–	+	+	+Δ	+	+

Contd...

Table 2–*Contd...*

Sl.No.	*Name of Grasshoppers*	*Jan*	*Feb.*	*March*	*April*	*May*	*June*	*July*	*Aug*	*Sept*	*Oct.*	*Nov.*	*Dec.*
	Family Tettiigoniidae												
	Subfamily–Conocephalinae												
43.	*C. maculates*	+	+0	–	–	–	–	–	+	+	+Δ	+	+
	Subfamily–Copiphorinae												
44.	*E. incertus* Walker	+	+0	–	–	–	–	–	+	+	+Δ	+	–
	Subfamily–Listroscelidinae												
45.	*H. unicolar* Serville	+0	–	–	–	–	–	+	+	+	+Δ	+	+
	Subfamily–Mecopodinae												
46.	*M. elongata* (Linnaeus)	+0	–	–	–	–	–	+	+	+	+Δ	+	+
	Subfamily–Phnaeropterinae												
47.	*L. intermedia* Ingrich	–	–	–	–	–	+	+	+Δ	+	+	+	+0
48.	*D. japonica* (Thunberg)	–	–	–	–	–	–	+	+Δ	+	+	+	+0
49.	*Tapiena latifolia** Ingrisch and shishodia	–	–	–	–	–	+	+	+	+	+	+	+
	Subfamily–Oxylakina												
50.	*Kirkaldyas truncate** (Karny)	–	–	–	–	–	–	+	+	+	+	+	+
	Subfamily–Pseudophyllinae												
51.	*Sathrophylla rugosa** L.	–	–	–	–	–	–	+	+	+	+	+	+

+: Indicates that species presence; –: indicates that species absence; Δ: indicates that maximum population; 0: indicates that minimum population; *: indicates that rare species.

August to February. From Family Tettiigonidae 6 species were reported. Out of which *D. japonica, H. unicolar* and *L. intermedia* were reported in the months July to January, *C. maculates* and *E. insertus* reported from July to February and *M. elongata* was reported from July to January.

The abundance of species found in Ghats region is recorded in Table 2. In all, 51 species of grasshoppers were reported belonging to four families, 21 subfamilies and 47 genera. Out of which 45 species were common and 7 species were rare. The family Acrididae contains three rare species *viz., I. convergences, D. venusta* and *A. flavescence* and family Tetrigidae contain only one rare species namely *T. bipunctata* and family Tettiigoniidae contain three rare species *viz., T. latifolia, K. truncate* and *S. rugosa.* All rare species of Ghats region are lacking in plain region. Seasonal abundance of most of families were mentioned in plain region which have same pattern of distribution found in Ghats region except, *P. infumata, G. semipictus* and *C. robustus* species belonging to family Acrididae were lacking.

Results

Results recorded in Table 3 shows that 51 species of grasshoppers from four families reported from Kolhapur region (Table 2) 27 species of family Acrididae, 4 species of family Pyrgomorphidae, 11 species of family Tetrigidae and 9 species of family Tettiigonidae were reported while, *A. thalassinus tamulus, O. hyla hyla, C. pinguis, S. splendens, X. humalis humalis, S. prasiniferum prasiniferum, G. laticornis, E. alacris alacris* species of family Acrididae were reported from all tahsils of Kolhapur district. The *A. crenulata* is only one species of family Pyrgomorphidae reported from all tahsils of Kolhapur district. However, *S. harpago, E. dorsifer, H. gracilis* and *H. linefera* of family Tetrigidae reported from all tahsils of Kolhapur district and from family Tettiigoniidae, *C. maculates, M. elongate, D. japonica, L. intermedia* and *H. unicolar* found in all tahsils of Kolhapur district *P. sunccinta, H. banian, A. flavecens, T. varicornis, H. bilineatus, E. tarsallis, C. sphenariodies, C. maculates, E. personatus, S. rugosa, K. truncata, T. latifolia,* all these species have been adapted for specific climatic conditions and reported only in particular habitat.

Grasshopper Species Distribution

Table 3: Distributional Patterns of Grasshoppers from Kolhapur District

Sl.No.	*Generic Name*	*Karvir*	*Kagal*	*Pan-hala*	*Gadhin-glaj*	*Chand-gad*	*Ajra*	*Bhud-argad*	*Radha-nagari*	*Shahu-wadi*	*Hatka-angale*	*Gagan-bavada*
	Family–Acrididae											
	Subfamily–Acridinae											
1.	*T. indica* Bol.	+	+	+	+	+	+	+	+	+	+	+
2.	*A. exaltata* Walker	+	+	+	+	+	+	+	+	+	+	+
3.	*P. infumata* Brunner	+	+	–	–	–	–	–	–	–	–	–
	Subfamily–Gomphocerinae											
4.	*G. laticornis* (Serv)	+	+	+	+	+	+	+	+	+	+	+
5.	*G. simipictus Walk.*	–	–	–	–	–	–	+	+	+	–	–
6.	*I. convergecne* (Walk)	+	+	–	–	–	–	–	–	–	–	–
	Subfamily–Oedipodinae											
7.	*D. venusta* (Walk.)	–	–	–	–	–	–	+	+	+	–	–
8.	*D. desius* walk	+	+	+	–	–	–	–	+	+	+	–
9.	*S. indica* sp.nov.	+	+	+	–	–	–	–	+	+	+	–
10.	*A. thalassinus tamulus* (Fabr.)	+	+	+	+	+	+	+	+	+	+	+
11.	*G. africanus africanus* Sauss.	+	+	+	+	+	+	+	+	+	+	+
12.	*H. respondens* (Walk)	+	+	+	+	+	+	–	+	+	+	–
	P. sahyadricus sp. nov.	+	+	+	+	+	+	+	+	+	+	+
13.	*T. annulata* (Th.)	+	+	+	+	+	+	+	+	+	+	+

Contd...

Table 3–*Contd...*

Sl.No.	*Generic Name*	*Karvir*	*Kagal*	*Pan-hala*	*Gadhin-glaj*	*Chand-gad*	*Ajra*	*Bhud-argad*	*Radha-nagari*	*Shahu-wadi*	*Hatka-angale*	*Gagan-bavada*
	Subfamily Hemiacridinae											
14.	*S. prasiniferum prasiniferum* (walk.)	+	+	+	+	+	+	+	+	+	+	+
15.	*H. banian*	+	+	+	–	–	–	–	–	–	–	–
16.	*H. bilineatus* Kirby	+	–	+	–	–	–	–	+	–	–	–
	Subfamily–Oxynae											
17.	*O. hyla hyla* (Serv.)	+	+	+	+	+	+	+	+	+	+	+
18.	*O. japonica japonica* (Th.)	+	+	+	–	+	–	–	+	+	+	+
	Subfamily–Cyrtacanthacridinae											
19.	*P. sunccincta* (Johonsson)	+	+	–	–	–	+	+	+	+	+	–
20.	*C. tatarica* (Linn.)	+	+	+	+	–	–	–	+	+	+	–
21.	*A. flavescence* (Fab.)	+	–	–	+	+	+	–	–	+	–	–
	Subfamily–Romalinae											
22.	*T. monticollis Gray*	+	–	–	+	+	+	+	+	+	+	–
	Subfamily–Catantopinae											
23.	*C. innotibilis* (Walk)	+	+	+	+	+	+	+	+	+	+	+
24.	*X. humalis humails* (Serv.)	+	+	+	+	+	+	+	+	+	+	+
25.	*S. splendens* (Th.)	+	+	+	+	+	+	+	+	+	+	+
	Subfamily–Eprepocenemidinae											
26.	*T. varicornis* (walk)	+	–	–	+	+	–	–	–	+	–	+
27.	*E. alacris alacris* (Serve)	+	+	+	+	+	+	+	+	+	+	+

Contd...

Table 3–*Contd...*

Sl.No.	*Generic Name*	*Karvir*	*Kagal*	*Pan-hala*	*Gadhin-glaj*	*Chand-gad*	*Ajra*	*Bhud-argad*	*Radha-nagari*	*Shahu-wadi*	*Hatka-angale*	*Gagan-bavada*
	Family Pyrgomorphidae											
	Subfamily–Orthacridinae											
28.	*C. sphenarioidides* Bolivar	+	+	–	–	–	–	–	–	+	+	–
	Subfamily–Pyrgomorphinae											
29.	*C. oxypterus* Blanchard	+	+	+	+	+	–	–	–	+	–	–
30.	*A. crenulata* (Fabricius)	+	+	+	+	+	+	+	+	+	+	+
31.	*T. apludi* sp. nov.	+	+	+	–	–	–	–	+	+	+	–
	Family–Tetrigidae											
	Subfamily–Scelimeninae											
32	*S. harpago* Serv.	+	+	+	+	+	+	+	+	+	+	+
33.	*T. pruthii* Gunther	+	+	+	+	+	–	–	+	+	–	+
34.	*A. cuneatus* Hanck.	+	+	–	–	+	–	+	–	+	–	+
	Subfamily–Metrodorinae											
35.	*S. cinereus* Brunn.	+	+	+	–	–	–	–	+	+	–	+
	Subfamily–Tetriginae											
36.	*T. bipunctata* (Linnaeus)	+	+	+	+	+	+	–	–	–	+	+
37.	*E. personatus* Bol.	+	+	–	–	–	+	–	–	–	+	+
38.	*H. gracilis* De Haan.	+	+	+	+	+	+	+	+	+	+	+
39.	*H. lineifera* Walk.	+	+	+	+	+	+	+	+	+	+	+

Contd...

Table 3–*Contd...*

Sl.No.	Generic Name	Karvir	Kagal	Pan-hala	Gadhin-glaj	Chand-gad	Ajra	Bhud-argad	Radha-nagari	Shahu-wadi	Hatka-angale	Gagan-bavada
40.	*E. dorsifer* (Walk)	+	+	+	+	+	+	+	+	+	+	+
41.	*E. tarsalis* Kirbay	+	–	–	+	+	+	+	+	+	+	–
42.	*E. guntheiri* Steninmann	+	+	+	+	+	+	+	+	+	+	+
	Family Tettiigonidae											
	Subfamily–Conocephalinae											
43	*C. maculates*	+	+	+	+	+	+	+	+	+	+	+
	Subfamily–Copiphorinae											
44	*E. insertus* Walk	+	+	–	–	–	+	+	+	+	+	+
	Subfamily–Listroscelidinae											
45.	*H. unicolar* Serville	+	+	+	–	+	+	+	+	+	+	+
	Subfamily–Mecopodinae											
46.	*M. elongata* (Linnaeus)	+	+	+	+	+	+	+	+	+	+	+
	Subfamily–Phnaeropterinae											
47.	*L. intermedia* Ingrich	+	–	+	+	+	+	+	+	+	+	+
48	*D. japonica* (Thunberg)	+	+	+	+	+	+	+	+	+	+	+
49	*T. latifolia* Ingrisch and Shishodia	+	–	–	–	+	+	–	–	–	–	–
	Subfamily–Oxylakina											
50	*K. truncate* (Karny)	+	–	–	–	+	+	–	–	–	–	–
	Subfamily–Pseudophyllinae											
51.	*S. rugosa* L.	+	+	–	–	–	–	–	+	+	–	–

Table 4: Distributional Patterns of Grasshoppers from Sangli District

Sl.No.	*Generic Name*	*Miraj*	*B. shirala*	*Islampur*	*Tasgaon*	*Jath*	*Chandoli*	*Sagresh-war*	*Vita*
	Family–Acrididae								
	Subfamily–Acridinae								
1.	*T. indica* Bol.	–	–	–	–	–	+	+	–
2.	*A. exaltata* Walker	+	+	+	+	+	+	+	–
3.	*P. infumata* Brunner	–	–	–	+	+	–	–	+
	Subfamily–Gomphocerinae								
4.	*G. laticornis* (Serv.)	–	+	+	–	–	+	+	–
	Subfamily–Oedipodinae								
5.	*D. venusta* (Walk.)	–	–	–	–	–	+	+	–
6.	*D. decisus* W.	+	+	+	+	+	+	+	+
	S. indica sp. nov.	–	–	–	–	–	+	+	–
7.	*A. thalassinus tamulus* (Fabr.)	+	+	+	+	+	+	+	+
8.	*G. africanus africanus* Sauss.	+	+	+	+	+	+	+	+
9.	*H. respondens* (Walk)	+	+	+	+	+	+	+	+
10	*P. sahydricus* sp. nov.	+	+	+	–	–	+	+	–
11.	*T. annulata* (Th.)	+	+	+	+	+	+	+	+
	Subfamily Hemiacridinae								
12.	*S. prasiniferum prasiniferum* Walk.	+	+	+	+	+	+	+	+
13.	*H. kolhapurensis* (Faber.)	–	+	+	–	–	+	+	–

Contd...

Table 4–*Contd...*

Sl.No.	*Generic Name*	*Miraj*	*B. shirala*	*Islampur*	*Tasgaon*	*Jath*	*Chandoli*	*Sagresh-war*	*Vita*
	Subfamily Oxynae								
14.	*O. hyla hyla* (Serv.)	+	+	+	+	+	+	+	+
15.	*O. japonica japonica* (Th.)	+	+	+	–	–	+	+	–
	Subfamily–Cyrtacanthacridinae								
16.	*P. sunccincta* (Johonsson)	–	–	–	–	–	+	+	–
17.	*C. tatarica* (Linn.)	+	–	–	–	+	+	+	+
18.	*A. flavescence* (Fab.)	–	–	–	–	–	+	+	–
	Subfamily–Romalinae								
19.	*T. monticollis* Gray	–	–	–	–	–	+	+	–
	Subfamily–Catantopinae								
20.	*C. innotibilis* (Walk)	+	+	+	+	+	+	+	+
21.	*X. humalis humails* (Serv.)	+	+	+	+	+	+	+	+
22.	*S. splendens* (Th.)	–	–	–	–	–	+	+	–
	Subfamily–Eprepocenemidinae								
23.	*T. varicornis* (walk)	–	–	–	–	–	+	+	–
24.	*E. alacris alacris* (Serve)	+	+	+	+	+	+	+	+
25.	*C. robustus* (Serville)	–	–	–	+	+	–	–	+

Contd...

Table 4–*Contd...*

Sl.No.	*Generic Name*	*Miraj*	*B. shirala*	*Islampur*	*Tasgaon*	*Jath*	*Chandoli*	*Sagresh-war*	*Vita*
	Family Pyrgomorphidae								
	Subfamily–Pyrgomorphinae								
26.	*C. oxypterus* Blanchard	–	–	–	+	+	+	+	–
27.	*A. crenulata* (Fabricius)	+	+	+	+	+	+	+	+
28.	*T. apludi* sp. nov.	–	+	+	–	–	+	+	–
	Family–Tetrigidae								
	Subfamily–Scelimeninae								
29.	*S. harpago* Serv.	+	+	+	+	+	+	+	+
30.	*T. pruthii* Gunther	+	+	+	+	+	+	+	+
31.	*A. cuneatus* Hanc.								
	Subfamily–Tetriginae								
32.	*T. bipunctata* (Linnaeus)	+	+	+	+	+	+	+	+
33.	*E. personatus* Bol.	+	+	+	–	–	+	+	–
34.	*H. grasilis* (De Hann)	+	+	+	–	–	+	+	–
35.	*H. lineifera* Walker	–	–	–	–	–	+	+	–
36.	*E. gunthery* Steinmann	+	+	+	+	+	+	+	+
37.	*E. dorsifer* (Walker)	+	+	+	–	–	+	+	–

Contd...

Table 4–*Contd...*

Sl.No.	*Generic Name*	*Miraj*	*B. shirala*	*Islampur*	*Tasgaon*	*Jath*	*Chandoli*	*Sagresh-war*	*Vita*
	Family Tettigoniidae								
	Subfamily–Conocephalinae								
38.	*C. maculates* (Le Guillou)	+	+	+	+	+	+	+	+
	Subfamily–Copiphorinae								
39.	*E. incertus* Walker	–	+	+	–	–	+	+	–
	Subfamily–Listroscelidinae								
40.	*H. unicolar* Serville	+	+	+	–	–	+	+	–
	Subfamily–Mecopodinae								
41.	*M. elongata* (Linnaeus)	+	+	+	–	–	+	+	
	Subfamily–Phnaeropterinae								
42.	*D. japonica* (Thunberg)	+	+	+	+	–	+	+	+
43.	*L. intermedia* Ingrich	+	+	+	–	–	+	+	–
	Subfamily–Pseudophyllinae								
44.	*S. rugosa* L.	–	–	–	–	–	+	–	–

Table 5: Distributional Patterns of Grasshoppers from Satara District

Sl.No.	*Generic Name*	*Patan*	*Sajjan-gad*	*Mahaba-leshwar*	*Karad*	*Wai*	*Lonand*	*Pachgani*
	Family–Acrididae							
	Subfamily–Acridinae							
1.	*T. indica* Bol.	+	+	+	+	+	+	+
2.	*A. exaltata* Walker	+	+	+	+	+	+	+
3.	*P. infumata* Brunner	+	+	–	+	–	–	–
	Subfamily–Gomphocerinae							
4.	*G. laticornis* (Serv)	+	+	+	+	+	+	+
	Subfamily–Oedipodinae							
5.	*D. desius* walk	+	+	+	+	+	+	+
6.	*S. indica* sp. nov.	–	+	+	–	+	–	–
7.	*A. thalassinus tamulus* (Fabr.)	+	+	+	+	+	+	+
8.	*G. africanus africanus* Sauss.	+	+	+	+	+	+	+
9.	*H. respondens* (Walk)	+	+	+	+	+	+	+
10.	*P. sahyadricus* sp. nov.	–	+	+	–	+	+	+
11.	*T. annulata* (Th.)	+	+	+	+	+	+	+
	Subfamily Hemiacridinae							
12.	*S. prasiniferum prasiniferum* (Walk.)	+	+	+	+	+	+	+
13.	*H. kolhapurensis* sp.nov.	–	+	+	+	+	–	–
	Subfamily–Oxynae							
14.	*O. hyla hyla* (Serv.)	+	+	+	+	+	+	+

Contd...

Table 5–*Contd...*

Sl.No.	*Generic Name*	*Patan*	*Sajjan-gad*	*Mahaba-leshwar*	*Karad*	*Wai*	*Lonand*	*Pachgani*
	Subfamily–Cyrtacanthacridinae							
15.	*P. succincta* (Johonsson)	+	+	+	+	+	+	+
16.	*C. tatarica* (Linn.)	–	+	–	+	–	–	–
	Sub family–Romalinae							
17.	*T. monticollis Gray*	–	–	+	+	+	+	+
	Subfamily–Catantopinae							
18.	*C. pinguise* (Walk)	+	+	+	+	+	+	+
19.	*X. humalis humails (*Serv.)	+	+	+	+	+	+	+
20.	*S. splendens* (Th.)	+	+	+	+	+	+	+
	Subfamily–Eprepocenemidinae							
21.	*E. alacris alacris* (serve)	+	+	+	+	+	+	+
22.	*T. varicornis* (walk)	–	+	–	+	+	–	–
	Family Pyrgomorphidae							
	Subfamily–Orthacridinae							
23.	*C. sphenarioidides* Bolivar	+	+	–	+	–	–	–
	Subfamily–Pyrgomorphinae							
24.	*C. oxypterus* Blanchard	+	+	–	+	–	–	–
25.	*A. crenulata* (Fabricius)	+	+	+	+	+	+	+

Contd...

Table 5–*Contd...*

Sl.No.	*Generic Name*	*Patan*	*Sajjan-gad*	*Mahaba-leshwar*	*Karad*	*Wai*	*Lonand*	*Pachgani*
	Family–Tetrigidae							
	Subfamily–Scelimeninae							
26.	*S. harpago* Serv.	+	+	+	+	+	+	+
27.	*A. cuneatus* Hanck.	+	+	–	+	+	–	–
28.	*T. purthi* Gunther	+	+	–	+	–	+	–
	Subfamily–Tetriginae							
29.	*E. personatus* Bol.	+	+	–	+	–	+	–
30.	*E. guntheri* Stenmann	+	+	+	+	+	+	+
31.	*H. gracilis* De Haan.	+	+	+	+	+	+	+
32.	*H. linfera* Walk.	+	+	+	+	+	+	+
33.	*E. dorsifer* (Walk)	+	+	–	+	–	–	–
34.	*E. tarsalis* Kirbay	+	+	–	+	+	–	–
	Family Tettigoniidae							
	Subfamily–Conocephalinae							
35.	*C. maculates* (Le Guillou)	+	+	+	+	+	+	+
	Sub Family–Copiphorinae							
36.	*E. incertus* Walker	+	+	+	+	+	+	+
	Subfamily–Listroscelidinae							
37.	*H. unicolar* Serville	+	+	+	+	+	+	+

Contd...

Table 5–*Contd...*

Sl.No.	*Generic Name*	*Patan*	*Sajjan-gad*	*Mahaba-leshwar*	*Karad*	*Wai*	*Lonand*	*Pachgani*
	Subfamily–Mecopodinae							
38.	*M. elongata* (Linnaeus)	+	+	+	+	+	+	+
	Subfamily–Phnaeropterinae							
39.	*D. japonica* (Thunberg	+	+	+	+	–	+	+
40.	*L. intermedia* Brumer	+	+	+	+	+	+	+
41.	**Subfamily–Pseudophyllinae**							
42.	*S. rugosa* (Linnaeus)	+	–	+	+	–	–	–

Results recorded in Table-4 shows that 25 species of family Acrididae were reported from Sangli district. Out of which 8 species of grasshopper were reported from all tahsils of Sangli district. Out of which *T. monticollis* and *A. flavescence* are rarely reported in Sangli district. *T. varicornis* is reported in good number from Chandoli and Sagareshwar region otherwise this species is not reported in other tahsils of district. Species distributinal studies was made from eight tahsils of Sangli district out of which Chandoli and Sagareshwar are very good spot for grasshopper diversity because, near about 22 species of grasshoppers have been reported from Sagareshwar and Chandoli, part of Western Ghats was available for very rich flora and fauna for grasshopper. Minimum diversity of grasshopper species reported in tahsils Tasgaon and Jath. The Tasgon and Jath region is dry, rainfall is very poor, and relatively host plant availablity is also very poor. Only well adapted species of grasshoppers are reported. The species population is also low as compare to Sagareshwar and Chandoli region. Miraj, Battis shirala and Islampur was with average population and diversity of grasshopper. Three members of Pyrgomorphidae also reported were namely *A. crenulata, C. oxypterus.* and *T. apludi* sp. nov. However, *A. crenulata* reported from all eight tahasils with good number but *C. oxypterus* reported only from Chandoli and Sagareshwar region.

Nine species of families Tetrigidae reported from Sangli district. All 9 species reported from eight tahsils of district, while, 7 species of Tettiigoniidae were reported in Sangli district. *C. maculates, and M. elongata,* reported from all eight tahsils of district. *C. robostus* is only species of family Acrididae reported from Sangli district.

Satara district and Kolhapur district's climatic conditions are more or less similar. Results recorded in Table 5 shows that 22 species of family Acrididae reported from Satara district. Out of which 15 species reported from 7 tahsils of district while, *D. venusta* reported in Sajjangad and Mahabaleshwar region. *C. tatarica* reported only in Sajjangad and Karad region. *A. flavescence, G. semipictus* and *C. robustus* species not reported in Satara district. Three member of family Pyrgomorphidae reported were *viz, A. crenulata, C. oxypterus* and *C. sphenarioides*. However, *A. crenulata* is more adopted species because they reported in all 7 region of Satara district. *C. oxypterus* reported in only Patan, Sajjangad and Karad region. while, *C. sphenarioides* reported only in Patan and Sajjangad region.

Family Tetrigidae also showed very rich diversity. *S. herpago, E. guntheri, H. gracilis* and *H. linifera* all are more adaptive as compare to other, they reported in all, 7 region of Satara district. Minimum population of Tetrigidae reported in Mahabaleshwar and Pachgani region. 7 species of family Tettiigonidae reported in 7 region of Satara district out of which *D. japonica, C. macculatus, E. incertus, M. elongata, H. unicolar,* and *L. intermedia* are more adapted because they reported in all seven regions of districts.

Results are recorded in Tables 6 to 31 and Figure 432 to 449 indicates that. 28 Grasshopper species from eight subfamilies namely, Acridinae, Gomphocerinae, Oedipodinae, Hemiacridinae, Oxynae, Cyrtacanthacridinae, Catantopinae and Eprepocenemidae belonging to two families Acrididae and Pyrgomorphidae from Plain region (Karveer) and Ghats region (Rdhanagri) are prevailing in the region of Kolhapur.

Four members of subfamily Acridinae reported in quadrate refer to *T. indica, A. exaltata, P. panteli,* and *P. infumata*. The species were reported in quadrate with good numbers in March to June from both Plain and Ghats region, Population of this species was high in Ghats region as compare to Plain region, *A. exaltata* was reported through out the year from Plain region with high population but in Ghats region it was reported only from April to December. Hence, population was poor in Ghats as compare to Plain region.

T. indica was reported in the quadrate from July to November, showing no much difference in population between Plain and Ghats region. However, *P. infumata* were reported only from plain region during the months July to November.

Three species of subfamily Gomphocerinae namely *G. laticirnis, G. semipictus* and *I. convergence* were reported under quadrate. *G. laticornis* reported from July to December in both Plain and Ghat region with a very high population while, *I. convergence* was found only in summer months with a low population. Very interestingly, *G. semipictus* rarely reported and that to from only Ghats region.

The subfamily Oedipodinae were more abundant as compare to other subfamilies, total six members were reported in quadrate. Out of which

Grasshopper Species Richness from Three Different Ecosystems of Karveer (2010-2011)

Ecosystems: a: Agricultural; b: Grassland; b: Semiforest.

Table 6

Name of Species	January			February			March			April			May			June		
	a	b	c	a	b	c	a	b	c	a	b	c	a	b	c	a	b	c
Sub family–Acridinae																		
T. indica	0	2	0	0	1	0	0	0	0	0	0	0	0	0	0	0	0	0
A. exaltata	0	1	0	0	0	1	0	0	2	1	1	0	1	0	0	1	0	0
P. panteli	0	0	0	0	0	0	0	4	0	0	7	0	0	6	0	0	6	0
P. infumata	0	0	0	0	0	0	0	0	0	0	0	0	0	0	0	0	0	0
Subfamily–Gomphocerinae																		
G. laticornis	0	0	0	0	0	0	0	0	0	0	0	0	0	0	0	0	0	0
I. convergence	0	0	0	0	0	0	0	2	0	4	0	0	2	0	0	0	0	0

July			August			September			October			November			December			T. No Speci-men	Sex Ratio	
a	B	c	a	b	c	a	b	c	a	b	c	a	b	c	a	b	c		Male	Female
0	0	0	1	0	0	2	1	1	2	2	1	1	1	1	1	1	0	15	6	9
1	0	0	0	2	0	1	2	1	2	2	1	1	0	1	0	0	2	24	14	10
0	0	0	0	0	0	0	0	0	0	0	0	0	0	0	0	0	0	23	11	12
1	0	1	1	0	2	0	0	2	1	0	0	0	0	1	0	0	0	9		
2	4	2	6	8	6	6	8	8	6	6	2	4	3	2	1	2	0	66	38	28
0	0	0	0	0	0	0	0	0	0	0	0	0	0	0	0	0	0	8	4	4

Table 7

Name of Species	January			February			March			April			May			June		
	a	*b*	*c*	*a*	*b*	*c*	*a*	*b*	*c*	*a*	*b*	*c*	*a*	*b*	*c*	*a*	*b*	*c*
Sub family–Oedipodinae																		
A. thalassinus tamulus	0	0	0	0	0	0	0	0	0	0	0	0	0	0	0	2	3	4
H. respondans	0	0	0	0	0	0	0	0	0	0	0	0	0	0	0	0	0	0
G. africanus africanus	0	3	1	0	4	0	2	4	0	1	3	0	1	2	0	2	2	0
T. annulata	1	2	2	2	3	1	2	2	1	2	3	0	2	3	3	6	3	2
D. decisus	0	0	0	0	0	0	0	0	0	0	0	0	0	0	0	0	0	0
Sub family Hemiacridinae																		
H. banian	0	0	0	0	0	0	0	0	0	0	0	0	0	0	0	0	0	0
P. bilineatus	0	0	0	0	0	0	0	0	0	0	0	0	0	0	0	0	0	0
S. prasiniferum prasiniferum	0	2	0	2	2	1	4	6	2	2	4	2	4	2	2	6	2	2

July			August			September			October			November			December			T. No Speci-men	Sex Ratio	
a	*B*	*c*	*a*	*b*	*c*	*a*	*b*	*c*	*a*	*b*	*c*	*a*	*b*	*c*	*a*	*b*	*c*		*Male*	*Female*
3	6	3	7	7	4	4	5	4	4	4	3	4	4	3	2	2	0	80	48	32
1	0	0	3	2	2	5	2	3	4	3	2	4	1	1	2	1	0	36	20	16
2	4	1	4	2	2	4	3	1	3	3	2	3	4	3	3	3	2	74	39	36
6	2	4	7	6	6	6	7	4	9	2	3	4	2	6	2	3	2	121	51	70
0	0	0	2	1	4	4	2	5	2	4	4	6	1	3	2	3	2	45	21	24
4	2	0	5	4	3	7	5	4	9	6	4	5	3	0	2	0	0	63	40	23
0	0	0	0	0	0	0	1	0	0	2	0	0	2	0	0	0	0	5	3	2
3	4	2	9	9	2	7	10	6	6	10	4	3	10	2	3	4	2	139	80	49

Table 8

Name of Species	January			February			March			April			May			June		
	a	b	c	a	b	c	a	b	c	a	b	c	a	b	c	a	b	c
Sub Family–Oxynae																		
O. hyla hyla	0	1	0	0	1	0	2	1	0	0	0	1	1	0	0	0	2	0
O. japonica japonica	0	0	0	0	0	0	0	0	0	0	0	0	0	0	0	0	0	0
Subfamily–Cyrtacanthacridinae																		
C. tatarica	0	0	0	1	0	0	0	0	0	0	0	0	0	0	0	0	0	0
P. suncincta	0	0	0	0	0	0	0	1	0	0	4	0	0	4	0	0	8	0
Subfamily–Catantopinae																		
C. pinguis innotabilis	0	0	0	0	0	0	0	0	0	0	0	0	2	2	1	2	3	1
S. splendens	0	0	0	0	0	0	0	0	0	0	0	0	0	0	0	2	1	0
X. humalis humalis	0	2	0	1	2	0	2	2	1	2	1	0	2	2	3	2	1	2

July			August			September			October			November			December			T. No Speci-men	Sex Ratio	
a	B	c	a	b	c	a	b	c	a	b	c	a	b	c	a	b	c		Male	Female
2	2	2	3	2	1	4	2	2	4	0	2	2	2	3	1	3	2	48	20	28
0	2	0	3	2	1	2	3	2	1	2	1	0	0	0	0	0	0	19	8	11
2	0	0	1	1	0	1	0	0	0	0	0	1	0	1	1	0	0	9	2	7
0	3	0	0	2	0	0	0	0	0	1	0	0	0	0	0	0	0	23	13	10
2	6	2	4	7	2	2	3	4	2	3	2	2	4	2	0	2	2	62	34	28
2	2	3	2	4	4	3	2	1	2	4	2	1	1	0	0	1	1	38	21	17
2	4	2	6	4	4	6	2	5	4	2	2	4	4	1	4	2	2	85	39	46

Table 9

Name of Species	January			February			March			April			May			June		
	a	b	c	a	b	c	a	b	c	a	b	c	a	b	c	a	b	c
Subfamily–Eprepocenemidinae																		
E. alacris alacris	0	0	0	0	0	0	0	0	0	0	2	0	0	2	0	2	4	2
T. varicornis	0	2	0	0	2	0	0	4	0	0	4	0	0	6	0	0	4	0
Family Pyrgomorphidae																		
C. oxypterus	0	2	0	0	1	0	0	0	0	0	0	0	0	0	0	0	2	1
A. crenulata	0	0	0	0	0	0	0	0	0	0	0	0	0	0	0	0	2	1
C. sphenarioides	0	0	0	0	0	0	0	0	0	0	0	0	0	0	0	0	0	0

July			August			September			October			November			December			T. No Speci-men	Sex Ratio	
a	B	c	a	b	c	a	b	c	a	b	c	a	b	c	a	b	c		Male	Female
2	3	1	4	6	2	2	6	3	1	3	1	2	3	2	0	1	0	54	26	27
0	0	0	0	0	0	0	0	0	0	0	0	0	0	0	0	0	0	22	13	9
1	2	0	0	4	0	1	5	0	1	4	0	0	2	0	0	2	0	28	12	16
2	6	2	6	7	2	4	4	2	2	4	4	2	3	4	1	2	2	62	32	30
0	7	0	0	9	0	0	20	0	0	26	0	0	9	0	0	3	0	36	20	16

Spimpson index: $D = \sum_{i=1}^{s} \frac{n_i(n_i - 1)}{N(N-1)}$

Where, n is the total number of organism of a particular species and N is the total number of organism of all species.

D = 0.95027 from Karveer region.

Grasshopper Species Richness from Three Different Ecosystems of Ghats Region (Radhanagari) (2010-2011)

Ecosystems: a: Agricultural; b: Platue; c: Semi forest

Table 10

Name of Species	January			February			March			April			May			June		
	a	b	c	a	b	c	a	b	c	a	b	c	a	b	c	a	b	c
Sub family–Acridinae																		
Truxalis indica Bol.	0	0	0	0	0	0	0	0	0	0	0	0	0	0	0	0	0	0
Acrida exaltata Walker	0	0	0	0	0	0	0	2	0	0	1	1	0	1	1	1	2	0
Phlaeoba panteli Bol.	0	0	0	0	0	0	3	2	0	2	2	0	2	6	3	9	10	6
Subfamily–Gomphocerinae																		
Gelostorrhinus laticornis (Serv.)	0	0	0	0	0	0	0	0	0	0	0	0	0	0	0	0	0	0
Gelostorrhinus semipectus (Walk)	0	0	0	0	0	0	0	0	0	0	0	0	0	0	0	0	0	0
Ischnacrida convergecne (Walk)	0	0	0	0	0	0	0	1	0	0	3	0	0	2	0	0	2	0

July			August			September			October			November			December			T. No Speci-men	Sex Ratio	
a	B	c	a	b	c	a	b	c	a	b	c	a	b	c	a	b	c		Male	Female
1	2	0	2	2	0	2	1	0	1	2	0	2	2	0	0	0	0	17	10	17
0	1	0	1	1	1	2	2	0	1	1	1	1	1	0	0	0	1	23	13	10
3	2	0	0	3	0	0	0	0	0	0	0	0	0	0	0	0	0	53	26	27
2	4	0	8	11	2	9	11	0	7	9	4	4	8	2	2	2	2	87	46	41
0	0	0	0	1	0	0	0	0	0	0	0	0	0	0	0	0	0	1	0	1
0	0	0	0	0	0	0	0	0	0	0	0	0	0	0	0	0	0	8	5	3

Table 11

Name of Species	January			February			March			April			May			June		
	a	*b*	*c*	*a*	*b*	*c*	*a*	*b*	*c*	*a*	*b*	*c*	*a*	*b*	*c*	*a*	*b*	*c*
Subfamily–Oedipodinae																		
Aiolopus thalassinus tamulus (Fab.)	0	0	0	0	0	0	0	0	0	0	0	0	0	0	0	3	4	2
Heteropternis respondens (Walk)	0	0	0	0	0	0	0	0	0	0	0	0	0	0	0	0	0	0
*Dittopternis venusta** (Walk)	0	0	0	0	0	0	0	0	0	0	0	0	0	0	0	2	2	0
Gastrimargus africanus africanus Sauss.	2	0	1	3	2	1	1	1	2	0	1	1	7	3	2	4	5	1
Trilophidia annulata (Th.)	2	2	1	2	2	0	2	2	0	0	3	0	2	4	0	2	2	0
Dociostaurus desius walk	0	0	0	0	0	0	0	0	0	0	0	0	0	0	0	2	2	0
Subfamily–Hemiacridinae																		
Hieroglyphus banian (Faber)	0	0	0	0	0	0	0	0	0	0	0	0	0	0	0	0	0	0
*Parahieroglyphus bilineatus**	0	0	0	0	0	0	0	0	0	0	0	0	0	0	0	0	0	0
Spathosternum prasiniferum prasiniferum (walk.)	2	4	3	2	3	2	3	2	0	2	1	1	1	2	1	2	2	1

Contd...

Table 1–*Contd...*

Sl.No. Name of Grasshoppers Jan Feb. March April May June July Aug Sept Oct. Nov. Dec.

July			*August*			*September*			*October*			*November*			*December*			*T. No Speci-men*	*Sex Ratio*	
a	*B*	*c*	*a*	*b*	*c*	*a*	*b*	*c*	*a*	*b*	*c*	*a*	*b*	*c*	*a*	*b*	*c*		*Male*	*Female*
7	6	2	7	4	2	8	5	3	6	4	2	4	4	0	2	2	0	77	39	28
0	1	0	3	3	2	4	1	1	1	2	0	0	2	0	0	1	0	21	10	11
2	1	0	0	0	0	0	0	0	0	0	0	0	0	0	0	0	0	7	4	3
2	7	3	2	3	4	1	4	2	3	5	1	4	6	2	2	3	1	89	50	39
2	4	0	4	4	0	4	3	0	1	2	0	2	1	0	1	0	0	54	28	26
1	2	4	4	4	2	4	2	2	2	1	1	1	2	1	1	2	1	41	26	15
1	0	0	2	0	0	5	3	0	4	2	0	2	1	0	0	0	0	20	13	7
0	0	0	0	0	0	2	2	0	1	2	0	0	0	0	0	0	0	7	3	4
7	2	6	6	6	8	9	11	12	6	2	2	4	4	3	3	2	2	129	60	69

Table 12

Name of Species	January			February			March			April			May			June		
	a	b	c	a	b	c	a	b	c	a	b	c	a	b	c	a	b	c
Sub Family–Oxynae																		
Oxya hyla hyla (Serv.)	0	0	1	0	2	0	0	0	0	0	1	0	0	0	0	0	0	0
Oxya japonica japonica (Th.)	0	0	0	0	0	0	0	0	0	0	0	0	0	0	0	0	0	0
Subfamily–Cyrtacanthacridinae																		
Cyrtacanthacris tatarica (Linn.)	0	0	0	0	0	0	0	2	0	0	0	0	0	1	0	0	0	0
Patanga succincta (Johonsson)	0	0	0	0	2	1	0	2	0	0	2	0	0	2	0	0	3	2
Subfamily–Catantopinae																		
Catantops innotibilis (Walk)	0	0	0	0	0	0	0	0	0	0	0	0	0	1	0	0	0	1
Stenocatantops splendens (Th.)	0	0	0	0	0	0	0	0	0	0	0	0	0	1	1	1	1	1
Xenocatantops humalis humails (Serv.)	2	4	0	2	3	0	2	2	0	1	2	0	1	3	0	2	4	0

July			August			September			October			November			December			T. No Speci-men	Sex Ratio	
a	B	c	a	b	c	a	b	c	a	b	c	a	b	c	a	b	c		Male	Female
0	2	0	1	1	0	2	3	1	1	0	1	0	0	3	0	2	1	22	10	12
1	0	0	2	1	0	2	2	0	2	1	0	0	0	0	0	0	0	11	4	6
0	0	0	0	1	0	0	1	1	0	0	0	0	0	0	0	0	0	6	2	4
0	3	0	0	0	0	0	0	0	0	0	0	0	0	0	0	0	0	17	9	8
2	1	0	1	1	3	1	0	1	2	0	1	0	0	1	0	1	0	17	10	6
0	1	1	1	1	2	1	1	2	1	2	0	0	0	1	0	0	1	20	11	9
3	6	0	4	4	0	4	6	0	2	4	0	3	2	0	1	2	0	69	32	37

Table 13

Name of Species	*January*			*February*			*March*			*April*			*May*			*June*		
	a	*b*	*c*	*a*	*b*	*c*	*a*	*b*	*c*	*a*	*b*	*c*	*a*	*b*	*c*	*a*	*b*	*c*
Subfamily–Eprepocenemidinae																		
Eyprepocenemis alacris alacris (Serve.)	0	0	0	0	0	0	0	0	0	2	2	0	1	3	2	2	4	4
Tylotropidius varicornis (walk)	0	0	0	0	6	3	0	6	4	0	6	5	0	7	4	0	5	7
Family Pyrgomorphidae																		
Chrotogonus oxypterus Blanchard	0	0	0	0	0	0	0	0	0	0	0	0	0	0	0	0	0	1
Atractomorpha crenulata (Fabricius)	0	0	0	0	0	0	0	0	0	0	0	0	0	0	0	0	0	0
Colemania sphenarioidides Bolivar	0	0	0	0	0	0	0	0	0	0	0	0	0	0	0	0	3	0

July			*August*			*September*			*October*			*November*			*December*			*T. No Speci-men*	*Sex Ratio*	
a	*B*	*c*	*a*	*b*	*c*	*a*	*b*	*c*	*a*	*b*	*c*	*a*	*b*	*c*	*a*	*b*	*c*		*Male*	*Female*
2	4	3	4	2	1	2	2	1	1	2	0	1	1	1	0	0	0	47	19	28
0	3	4	0	0	0	0	0	0	0	0	0	0	0	0	0	0	0	60	36	24
2	1	0	0	2	0	0	1	0	1	2	0	0	2	0	0	1	0	13	6	7
3	4	3	3	5	2	5	7	3	3	5	2	4	2	3	1	3	1	59	28	31
0	7	0	0	9	0	0	9	0	0	8	0	0	2	0	0	2	0	40	23	17

Spimpson index: $D=\sum_{i=1}^{s}\frac{n_i(n_i-1)}{N(N-1)}$

Where, n is the total number of organism of a particular species and N is the total number of organism of all species.

D = 0.941801 from Radhanagari region

Table 14: Average Population of Subfamily Acridinae and Truxalinae from Three Ecosystems of Plain Region (Karveer)

Name of Species	*Jan*	*Feb*	*Mar*	*Apr*	*May*	*Jun*	*Jul*	*Aug*	*Sep*	*Oct*	*Nov*	*Dec*
T. indica	0.66	0.33	0	0	0	0	0.66	1.66	1.33	0.33	1	0
A. exaltata	0.33	0.66	0.66	0.33	0.33	0.33	0.33	0.66	1.66	1.33	0.66	0.66
P. panteli	0	0	1	2	2	2.33	0.33	0	0	0	0	0
P. infumata	0	0	0	0	0	0	0.66	0.33	0.66	1	0.33	0

Table 15: Average Population of Subfamily Acridinae and Truxalinae from Three Ecosystems of Ghat Region (Radhanagari)

Name of Species	*Jan*	*Feb*	*Mar*	*Apr*	*May*	*Jun*	*Jul*	*Aug*	*Sep*	*Oct*	*Nov*	*Dec*
T. indica	0	0	0	0	0	0	1	1.33	1	1	1.33	0
A. exaltata	0	0	0	0.66	0.66	1	0.33	1	1.33	1	0.66	0.33
P. panteli	0	0	1.66	1.33	3.66	8.33	1.66	1	0	0	0	0

Table 16: Average Population of Subfamily Gomphocerinae from Three Ecosystems of Plain Region (Karveer)

Name of Species	*Jan*	*Feb*	*Mar*	*Apr*	*May*	*Jun*	*Jul*	*Aug*	*Sep*	*Oct*	*Nov*	*Dec*
G. laticornis	0	0	0	0	0	0	2.66	6.66	7.33	4.66	3	1
G. semipictus	0	0	0	0	0	0	0	0	0	0	0	0
I. convergences	0	0	0.33	0.33	0.33	0	0	0	0	0	0	0

Table 17: Average Population of Subfamily Gomphocerinae from Three Ecosystems of Ghat Region (Radhanagari)

Name of Species	*Jan*	*Feb*	*Mar*	*Apr*	*May*	*Jun*	*Jul*	*Aug*	*Sep*	*Oct*	*Nov*	*Dec*
G. laticornis	0	0	0	0	0	0	2	7	6.66	6.66	4.66	1.66
G. semipictus	0	0	0	0	0	0	0	0.33	0	0	0	0
I. convergens	0	0	0.33	1	0.66	0.66	0	0	0	0	0	0

Table 18: Average Population of Subfamily Oedipodinae from Three Ecosystems of Plain Region (Karveer)

Name of Species	*Jan*	*Feb*	*Mar*	*Apr*	*May*	*Jun*	*Jul*	*Aug*	*Sep*	*Oct*	*Nov*	*Dec*
H. respondes	0	0	0	0	0	0	0.33	3.22	2.33	3	2	1
G. a. africanus	1.33	1.33	2	1	1.33	1.33	2.33	2.66	2.66	2.66	3.33	2.66
A. thalassinus tamulus	0	0	0	0	0	2.66	4	4.33	6	4	3.66	1.33
T. annulata	1.66	2	1.66	1.66	2.66	3.66	4	6.33	5.66	4.66	4	2.33
D. desius	0	0	0	0	0	0	1	2.33	3.66	3.33	3.33	2.33

Table 19: Average Population of Subfamily Oedipodinae from Three Ecosystems of Ghat Region (Radhanagari)

Name of Species	*Jan*	*Feb*	*Mar*	*Apr*	*May*	*Jun*	*Jul*	*Aug*	*Sep*	*Oct*	*Nov*	*Dec*
H. respondes	0	0	0	0	0	0	0.33	2.66	2	1	0.66	0.33
G. a. africanus	1	2	1.33	0.66	4	3.33	2.33	3	2.33	3	4	2
A. thalassinus tamulus	0	0	0	0	0	3	5	4.33	5.33	4	2.33	1.33
T. annulata	1.66	1.33	1.33	1	2	1.33	2	2.66	2.33	1	1	0.33
D. desius	0	0	0	0	0	1.33	2.33	3.33	2.66	1.33	1.33	1.33
D. venusta	0	0	0	0	0	1.33	1	0	0	0	0	0

Table 20: Average Population of Subfamily Hemiacridinae from Three Ecosystems of Plain Region (Karveer)

Name of Species	*Jan*	*Feb*	*Mar*	*Apr*	*May*	*Jun*	*Jul*	*Aug*	*Sep*	*Oct*	*Nov*	*Dec*
H.banian	0	0	0	0	0	0	2	4	6.33	5.33	2.66	0.66
P. bilineatus	0	0	0	0	0	0	0	0	0.33	0.66	0.66	0
S. p. prasiniferum	2.66	1.66	4	0.66	2.66	3.33	3	6.33	7.66	6.66	5	2.66

Table 21: Average Population of Subfamily Hemiacridinae from Three Ecosystems of Ghat Region (Radhanagari)

Name of Species	*Jan*	*Feb*	*Mar*	*Apr*	*May*	*Jun*	*Jul*	*Aug*	*Sep*	*Oct*	*Nov*	*Dec*
H. kolhapurensis	0	0	0	0	0	0	0.33	0.66	2.66	2	1	0
P. bilineatus	0	0	0	0	0	0	0	0	1.33	1	0	0
S. p. prasiniferum	2.33	2.33	1.66	1.33	1.66	1.66	5	6.66	10.33	3.33	3.66	2.33

Table 22: Average Population of Subfamily Oxyinae from Three Ecosystems of Plain Region (Karveer)

Name of Species	*Jan*	*Feb*	*Mar*	*Apr*	*May*	*Jun*	*Jul*	*Aug*	*Sep*	*Oct*	*Nov*	*Dec*
O. hyla hyla	0.33	0.33	1	0.33	0.33	0.66	2	2	2.66	2	2.33	2
O. japonica japonica	0	0	0	0	0	0	0.66	2	2.33	1.33	0	0

Table 23: Average Population of Subfamily Oxyinae from Three Ecosystems of Ghat Region (Radhanagari)

Name of Species	*Jan*	*Feb*	*Mar*	*Apr*	*May*	*Jun*	*Jul*	*Aug*	*Sep*	*Oct*	*Nov*	*Dec*
O. hyla hyla	0.33	0.66	0.33		0	0	0.66	0.66	2	0.66	1	1
O. japonica japonica	0	0	0	0	0	0	0.33	1	1.33	1	0	0

Table 24: Average Population of Subfamily Cyrtacanthacridinae from Three Ecosystems of Plain Region (Karveer)

Name of Species	*Jan*	*Feb*	*Mar*	*Apr*	*May*	*Jun*	*Jul*	*Aug*	*Sep*	*Oct*	*Nov*	*Dec*
C. tatarica	0	0.33	0.66	0	0	0	0	0.66	0.33	0	0.66	0.33
P. suncincta	0	0	0.33	1.33	1.33	2.66	1	0.66	0	0.33	0	0

Table 25: Average Population of Subfamily Cyrtacanthacridinae from Three Ecosystems of Ghat Region (Radhanagar)

Name of Species	*Jan*	*Feb*	*Mar*	*Apr*	*May*	*Jun*	*Jul*	*Aug*	*Sep*	*Oct*	*Nov*	*Dec*
C. tatarica	0	0.33	0.66	0	0	0	0	0.33	0.66	0	0	0
P. suncincta	0	0	0.33	1.33	1.33	2.33	1	0.66	0	0.33	0	0

Table 26: Average Population of Subfamily Catantopinae from Three Ecosystems of Plain Region (Karveer)

Name of Species	*Jan*	*Feb*	*Mar*	*Apr*	*May*	*Jun*	*Jul*	*Aug*	*Sep*	*Oct*	*Nov*	*Dec*
C. pinguis innotabilis	0	0	0	0	1.66	2	3.33	4.33	3	2.33	2.66	1.33
C. splendens	0	0	0	0	0	1	2.33	3.33	2	2.66	0.66	0.66
X. humalis humalis	0.66	1	2.33	1	2.33	1.66	2.66	4.33	4.66	2.66	3	2.66

Table 27: Average Population of Subfamily Catantopinae from Three Ecosystems of Ghat Region (Radhanagari)

Name of Species	*Jan*	*Feb*	*Mar*	*Apr*	*May*	*Jun*	*Jul*	*Aug*	*Sep*	*Oct*	*Nov*	*Dec*
C. pinguis innotabilis	0	0	0	0	0.33	0.33	1	1.66	0.66	1	0.33	0.33
C. splendens	0	0	0	0	0.66	1	0.66	1.33	1.33	1	0.33	0.33
X. humalis humalis	2	1.66	1.33	1	1.33	2	3	2.66	3.33	2	1.66	1

Table 28: Average Population of Subfamily Eprepocenemidae from Three Ecosystems of Plain Region (Karveer)

Name of Species	*Jan*	*Feb*	*Mar*	*Apr*	*May*	*Jun*	*Jul*	*Aug*	*Sep*	*Oct*	*Nov*	*Dec*
E. alacris alacris	0	0	0	0.66	0.66	2.66	2	4	3.66	1.66	2.33	0.33
T. varicornis	0.33	0.33	1.33	1.33	2	2	0	0	0	0	0	0

Table 29: Average Population of Subfamily Eprepocenemidae from Three Ecosystems of Ghat Region (Radhanagari)

Name of Species	*Jan*	*Feb*	*Mar*	*Apr*	*May*	*Jun*	*Jul*	*Aug*	*Sep*	*Oct*	*Nov*	*Dec*
E. alacris alacris	0	0	0	1.33	2	3.33	3	2.33	1.66	1	1	0
T. varicornis	0	3	3.33	3.66	3.66	4	2.33	0	0	0	0	0

Table: 30: Average Population of Subfamily Pyrgomorphidae from Three Ecosystems of Plain Region (Karveer)

Name of Species	*Jan*	*Feb*	*Mar*	*Apr*	*May*	*Jun*	*Jul*	*Aug*	*Sep*	*Oct*	*Nov*	*Dec*
A. crenulata	0	0	0	0	0	1	3.33	3.33	5	3.33	3	1.66
C. oxypterus	0.66	0.33	0	0	0	1	1	1.33	2	1.66	0.66	0.66
C. sphenarioides	0	0	0	0	0	0	2.33	3	3	2.66	3	1

Table 31: Average Population of Subfamily Pyrgomorphidae from Three Ecosystems of Ghat Region (Radhanagari)

Name of Species	*Jan*	*Feb*	*Mar*	*Apr*	*May*	*Jun*	*Jul*	*Aug*	*Sep*	*Oct*	*Nov*	*Dec*
A. crenulata	0	0	0	0	0	0	3.66	4.33	5	3	2.66	1
C. oxypterus	0	0	0	0	0	1	2.33	3	3	2.66	0.66	0.66
C. sphenarioides	0	0	0	0	0	0	1	2.33	3	2.66	0.66	0.66

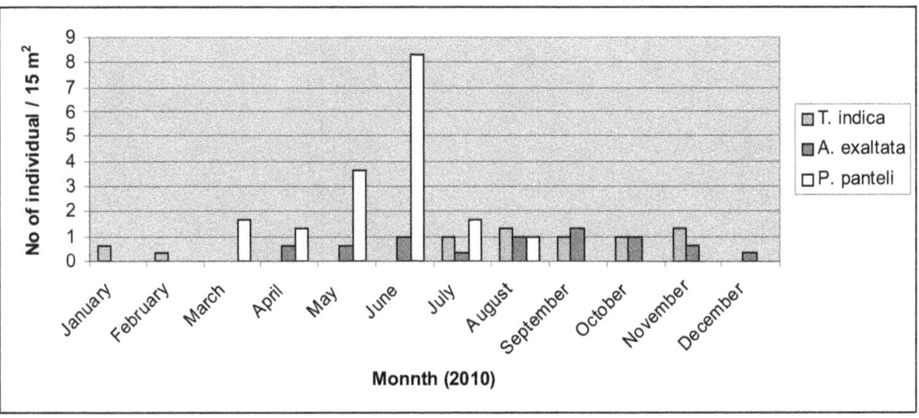

Figure 432: Density of Subfamily Acridinae from Plain Region (Karveer)

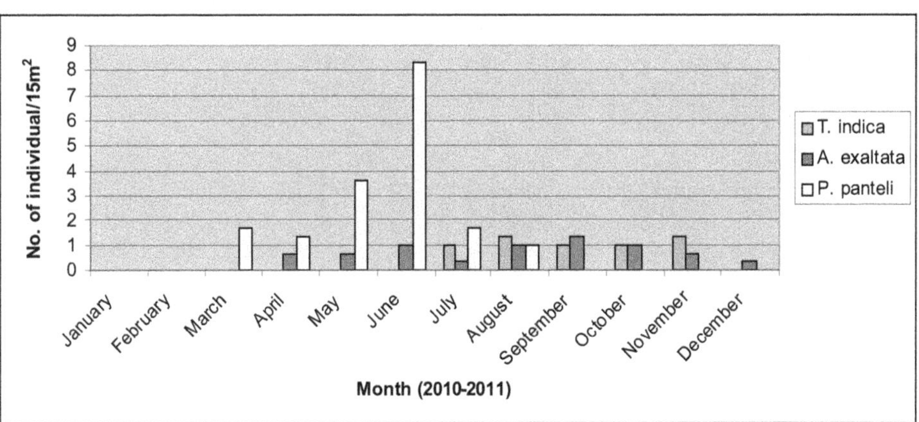

Figure 433: Density of Subfamily Acridinae from Ghats Region (Radhanagari)

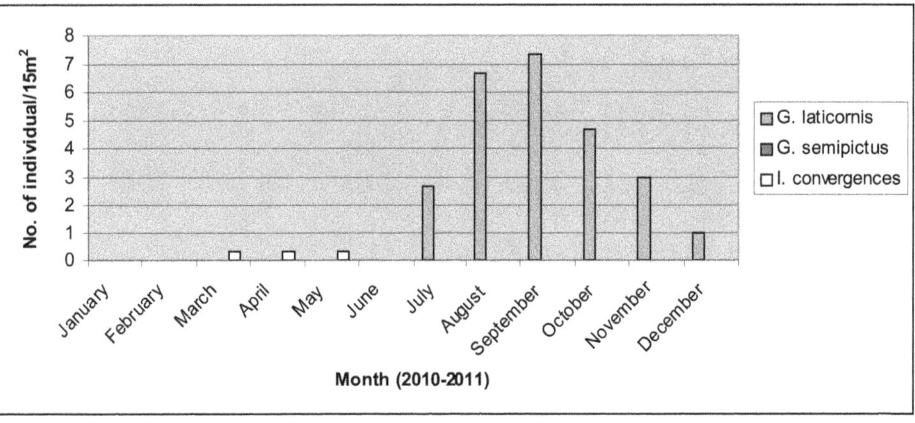

Figure 434: Density of Subfamily Gomphocerinae from Plain Region (Karveer)

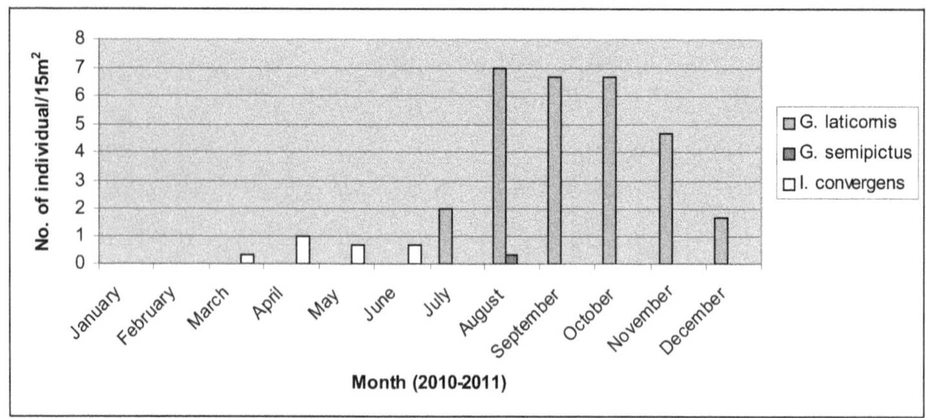

Figure 435: Density of Subfamily Gomphocerinae from Ghats Region (Radhanagari)

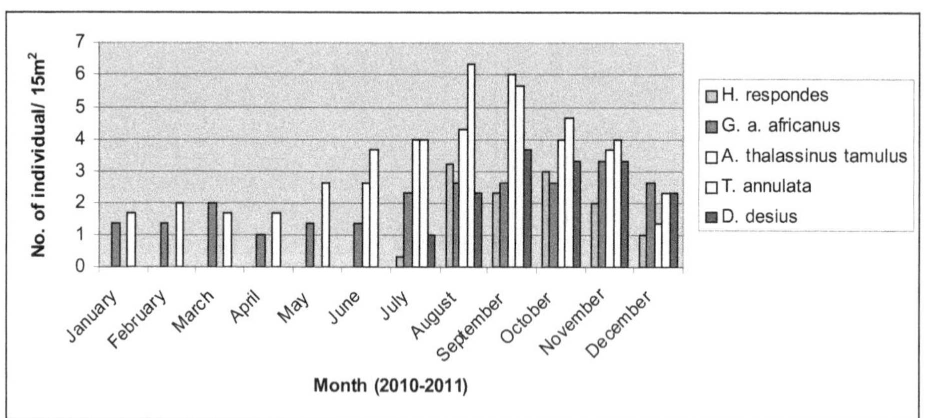

Figure 436: Density of Subfamily Oedipodinae from Plain Region (Karveer)

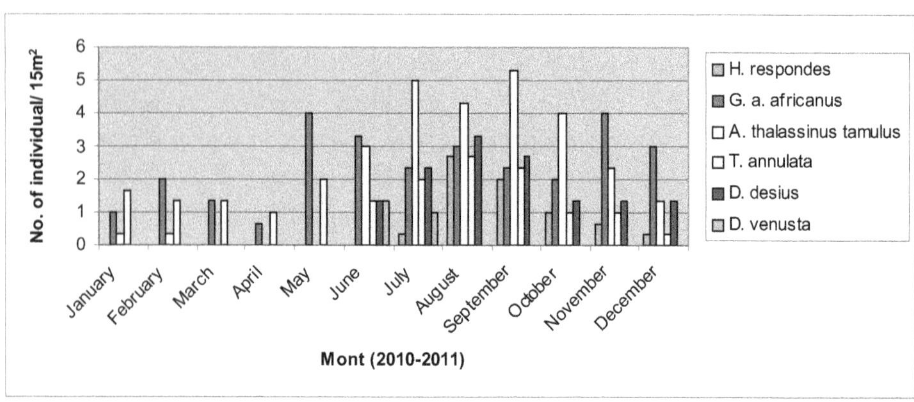

Figure 437: Density of Subfamily Oedipodinae from Ghats Region (Radhanagari)

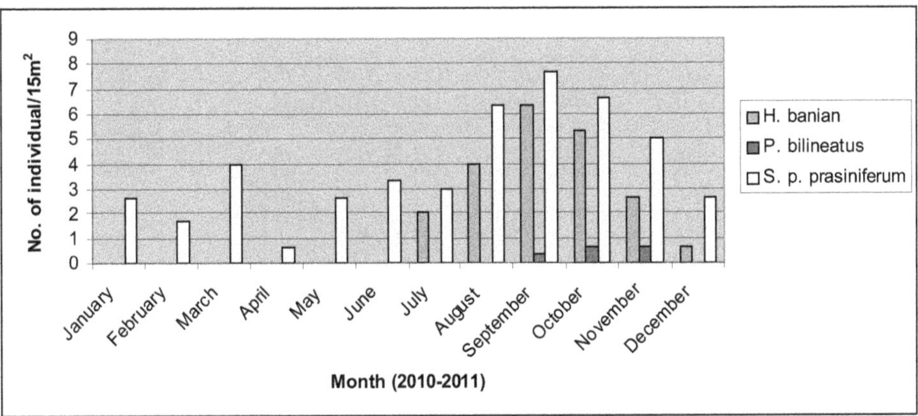

Figure 438: Density of Subfamily Hemiacridinae from Plain Region (Karveer)

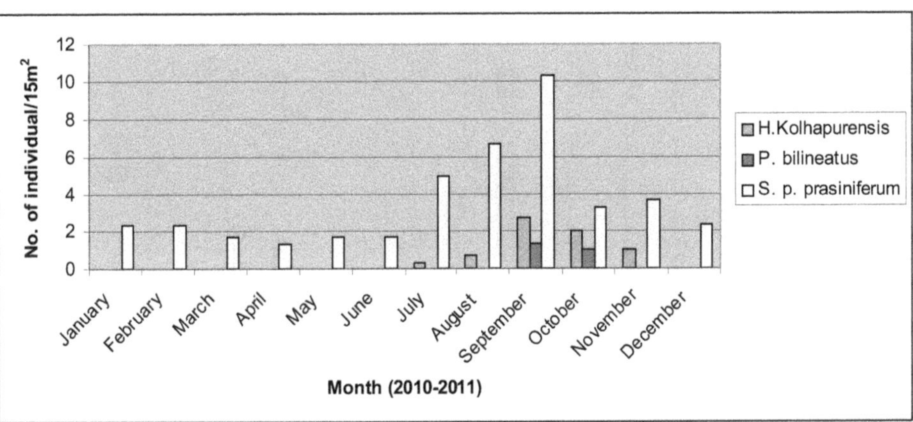

Figure 439: Density of Subfamily Hemiacridinae from Ghats Region (Radhanagari)

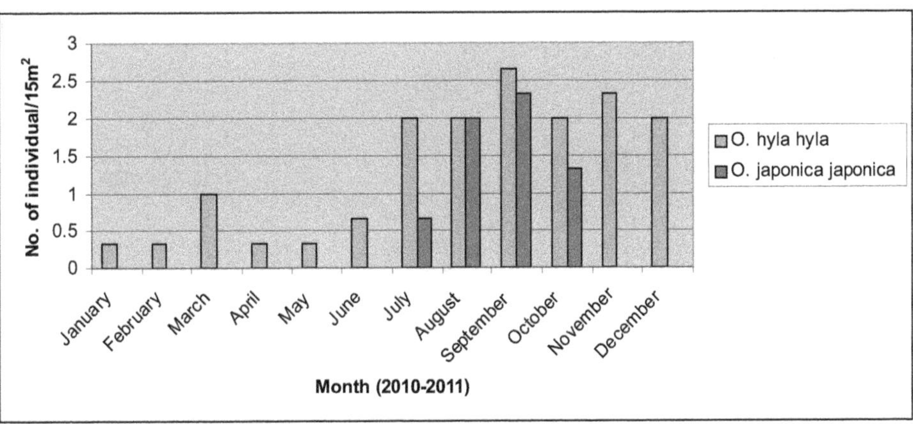

Figure 440: Density of Subfamily Oxyinae from Plain Region (Karveer)

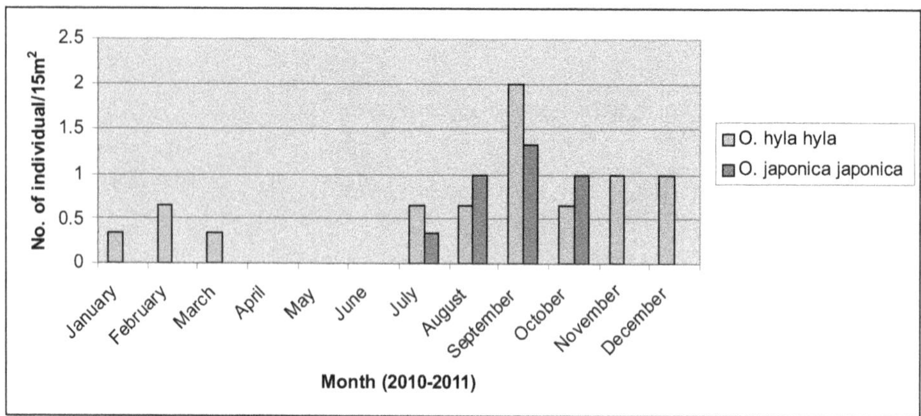

Figure 441: Density of Subfamily Oxyinae from Ghats Region (Radhanagari)

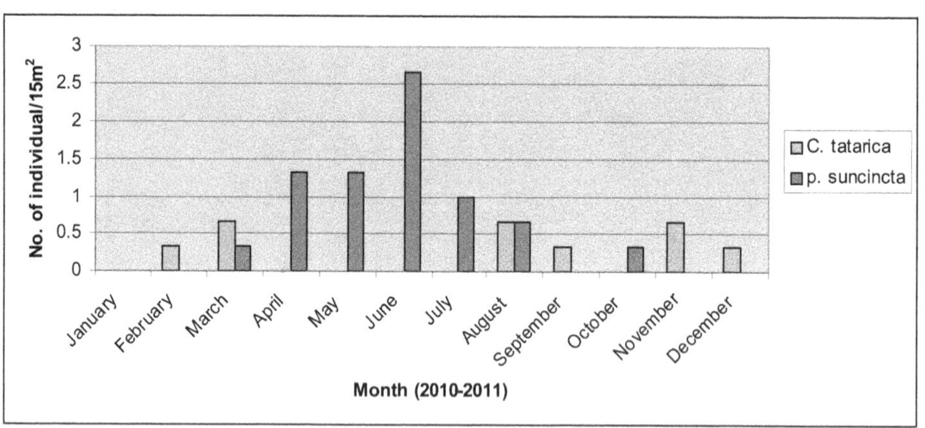

Figure 442: Density of Subfamily Cyrtacanthacridinae from Plain Region (Karveer)

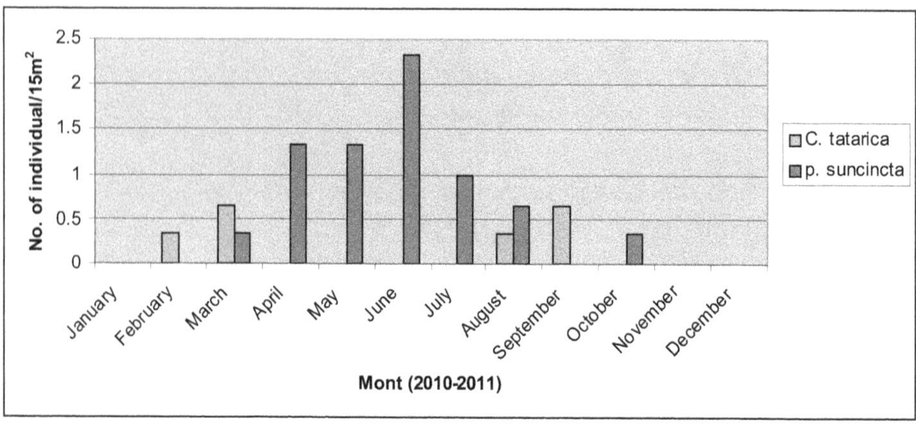

Figure 443: Density of Subfamily Cyrtacanthacridinae form Ghats Region (Radhanagari)

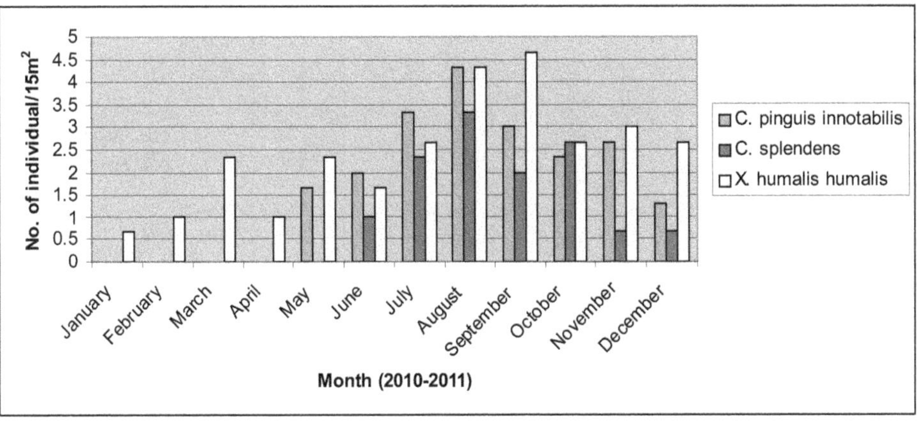

Figure 444: Density of Subfamily Catantopinae from Plain Region (Krveer)

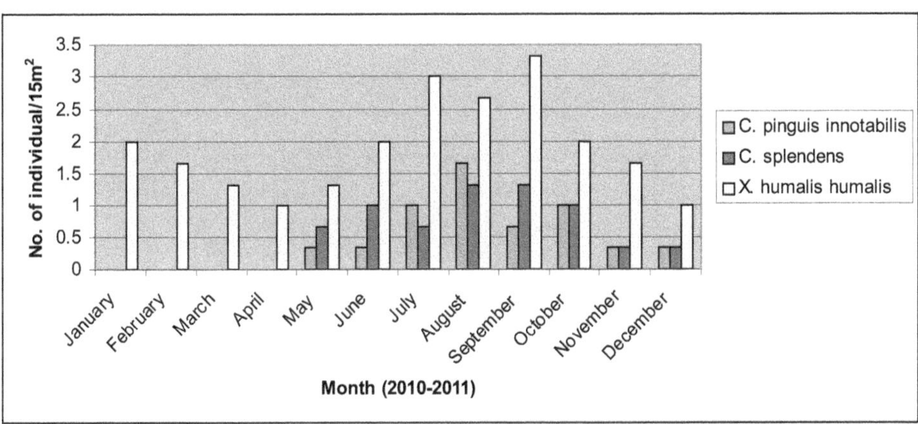

Figure 445: Density of Subfamily Catantopinae from Ghats Region (Radhanagari)

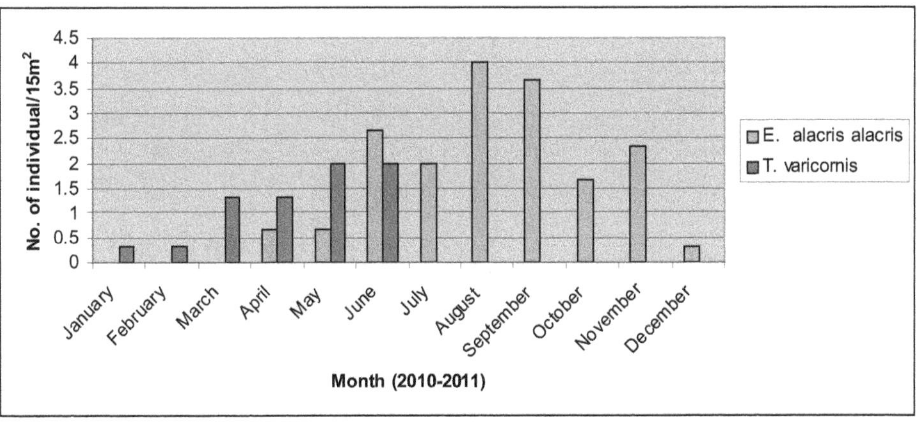

Figure 446: Density of Subfamily Eprepocenemidae from Plain Region (Karveer)

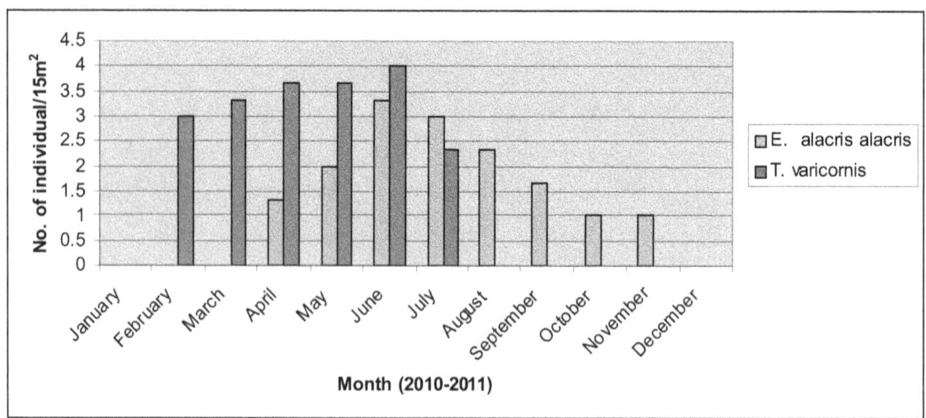

Figure 447: Density of Subfamily Eprepocenemidae from Ghats Region (Radhanagari)

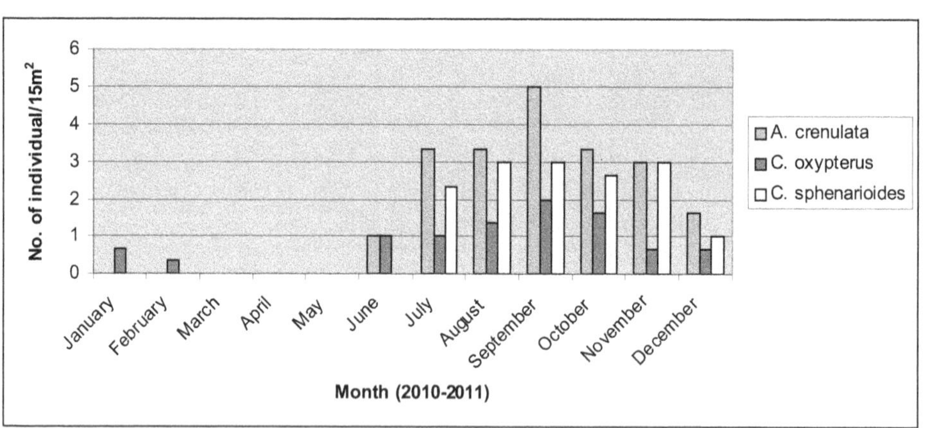

Figure 448: Density of Subfamily Pyrgomorphidae from Plain Region (Karveer)

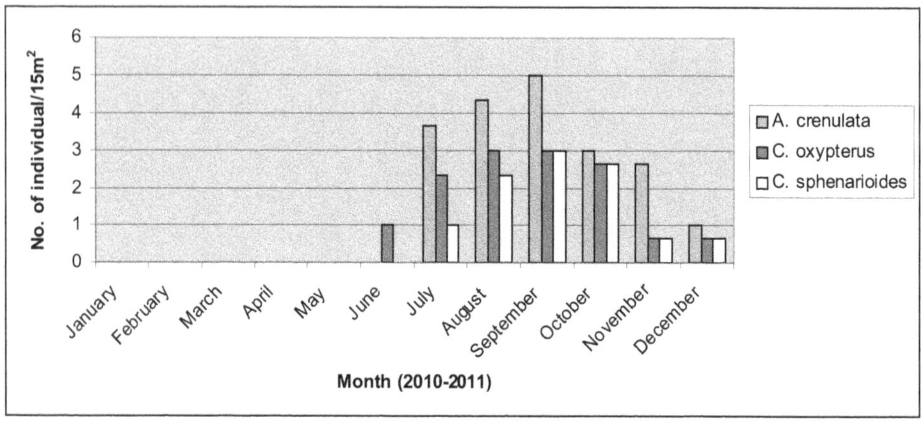

Figure 449: Density of Subfamily Pyrgomorphidae from Ghats Region (Radhanagari)

T. annulata, G. africanus africanus, A. thalasinus tamulus were dominant in both regions. The population of *T. annulata* was high in Plain region as compared to Ghats region. No much difference was found in the population of *G. africanus africanus* and *A. thalasinus tamulus* in Ghats and Plain region. *H. respondance* were also average population reported from both region during July to December while, *D. venusta* was found only in Ghats region with very low population.

The Subfamily Hemiacridinae was also important subfamily from the view point of species richness. Three members namely *S. prasiniferum prasiniferum, H. kolhapurensis* and *P. bilineatus*, reported in quadrate from both Plain and Ghats region. Out of which *S. prasiniferum prasiniferum* was more dominant in both regions. *H. Kolhapurensis* species population was high in plain region as compared to Ghats region. However, *P. bilineatus* was reported rarely in both regions.

Two members of subfamily Oxynae were reported in quadrate *viz. O. hyla hyla* and *O. japonica japonica*. The *O. hyla hyla* recorded with high number in both Plain and Ghats region during the months April to June. However, they were not reported from Ghats region and *O. japonica japonica* was reported from July to October in quadrate but with a very high population in Plain region. Similarly, two members (*C. tatrica* and *P. sunccincta*) of family Cyrtacanthacridinae were also reported in quadrate. *P. suncincta* dominantly reported from April to July in both the regions while *C. tatarica* was with a low population.

The Subfamily Catantopinae was also one of the richly reported subfamily from both Ghats and Plain regions. Three species (*C. pinguis, S. spledens* and *X. humalis humalis*) of this subfamily have been reported in quadrate. The species *X. humalis humalis* was quite dominant species in both regions. But the population of this species was comparatively high in Plain region. *C. pinguis* and *S. slendence* species were reported from May to December with high population in Plain region.

Two members of subfamily Eprepocenemidae namely. *E. alacris alacris* and *T. varicornis* were reported from Plain and Ghats region but *E. alacris alacris* was dominant in April to November, population peak was found in months June-July from Ghats region and August- September from Plain region. *T. varicornis* abundantly reported in months February to July, with maximum population from Ghats region.

Similarly, three members of family pyrgomorphidae namely *A. crenulata, C. oxypterus* and *C. spenarioides* were reported during June to February. *A. crenulata* was the dominant species from both Plain and Ghats region with good population while, C. *oxypterus* was reported abundantly in Ghats region and *C. sphenarioides* on an average reported in both Plain and Ghats region. Observations of distribution and seasonal abundance suggest that grasshoppers are major source of plant distruction both in Plain and Ghats region and there is need to maintain their population at economic level by keeping environment safe.

Discussion

Mayya *et al.* (2005) studied the abundance of 28 species of short horned grasshopper from Dakshina Kannada district of Karnatka. *A. exaltata, Stauradorus bicolar, Morphacris faciata, D. venusta, Oedulus abruptus* and *Acrotylus humbertianus* were relatively very much abundant than other species of family Acrididae. The family Acrididae was abundant during February and March while Pyrgomorphidae was present in large number during December and January.

Kandibane *et al.* (2004) studied the relative abundance of grasshoppers in irrigated rice ecosystem from Tamil Nadu, 21 species of grasshoppers were reported from protected and unprotected plots during the seasons Rabi 2000 and Kharif 2001. Two species of short horned grasshoppers *viz. Oxay nititulla* (Walk) and *O. fuscovitata* (Marsch) were dominant in both the ecosystem, also reported *Hieroglyphus banian* Fab. as the dominant species during Kharif 2001. Other short horned grasshoppers namely *A. exaltata. G. africanus africanus. Chrotogonus brachypterus* Bal., C. oxypterus, *Acrotylus humbertinus* Sass. *C. pimguise C. ferruginus* Bol., *E. alacris alacris, C. tatarica, A. flavisence and A. crenulata* were the rare species and among the long horned grasshopper species *Spenaroptera grasilus* Bur. *Conocephalus chinensis, C. maculates* and *Holoclora albida* Brun.were most common.

In the present work, abundance of families Acrididae, Tetrigidae, Pyrgomorphidae and Tettiigoniidae was studied from Plain and Ghats region of Kolhapur, Satara, and Sangli districts of Westerm Maharashtra.

Table 31: Check List of Grasshoppers from Three Districts (Satara, Sangli and Kolhapur) of Western Maharashtra including Ghats

Grasshopper Species	*Relative Host*		*Region*				
			a	*b*	*c*	*d*	*e*
FAMILY–ACRIDIDAE							
Subfamily–Acridinae							
Tribe–Truxalini							
1. *Truxalis indica* Bol.	Paddy	*Oryza sativa* L.	+	+	+	+	+
	Wheat	*Triticum aestivum* L.					
	Grass	*Apluda mutica* L.					
	Grass	*Heteropogon contortus* (L.)					
Tribe–Acridini							
2. *Acrida exaltata* Walk.	Maize	*Zea mays* L.	+	+	+	+	+
	Paddy	*O. sativa*					
	Grass	*A. mutica*					
	Wheat	*T. aestivum*					
Tribe–Phlaeobini							
3. *Phlaeoba infumata* Brunner.	Sugarcane	*Sacchaunum spontanum*	+	+	+	–	+
	Maize	*Z. mays*					
	Wheat	*T. aestivum* L					
4. *Phlaeoba panteli* Bol.	Maize	*Z. mays*	+	–	–	+	+
	Wheat	*T. aestivum*					
	Paddy	*O. sativa*					
	Grass	*H. contortus*					

Contd...

Table 31–*Contd...*

Grasshopper Species		*Relative Host*	*Region*				
			a	*b*	*c*	*d*	*e*
Subfamily–Gomphocerinae							
5. *Gelostorrhinus laticornis* (Serv.)	Paddy	*O. sativa*	+	+	+	+	+
	Grass	*H. contortus*					
	Grass	*T. quadrivalvis*					
6. *Gelostorrhinus semipectus* Walk.	–	–	+	–	–	–	+
7. *Ischnacrida convergence* (Walk)	Grass	*A. mutica* L.	+	–	–	+	+
	Grass	*H. contortus*					
	Grass	*T. quadrivis*					
Subfamily–Oedipodinae							
8. *Dittopternis venusta* (Walk)			+	–	+	–	+
9. *Dociostaurus desius* Walk	Wheat	*T. aestivum*	+	+	+	+	+
	Maize	*Z. mays*					
	Paddy	*O. sativa*					
	Grass	*H. contartus*					
	Grass	*T. quadrivalvis*					
10. *Stauradorus indica* sp. nov.	Grass	*A. mutica* L.	+	+	+	+	–
	Grass	*H. contortus*					
	Grass	*T. quadrivalvis*					

Contd...

Table 31–*Contd...*

Grasshopper Species	*Relative Host*		*Region*				
			a	*b*	*c*	*d*	*e*
Tribe–Epacromiini							
11. *Aiolopus thalassinus tamulus* (Fabr.)	Wheat	*T. aestivum*	+	+	+	+	+
	Maize	*Z. mays*					
	Tur	*C. cajan*					
	Paddy	*O. sativa*					
Tribe–Locustini							
12. *Gostrimargus africanus africanus* Sauss	Wheat	*T. aestivum*	+	+	+	+	+
	Maize	*Z. mays*					
	Geern gram	*V. radiata*					
	Tur	*C. cajan*					
	Paddy	*O. sativa*					
13. *Heteropternis respondens* (Walk)	Tur	*Cajanus cajan* L.Mill.	+	+	+	+	+
	Pea	*Pisum sativum* L.					
	Tomao	*L. esculentum*					
	Grass	*Themeda quadrivalvis* (L.)					
	Paddy	*O. sativa*					
14. *Peternoscirta sahyadricus* sp.nov.	Paddy	*O. sativa*	+	+	–	+	+
	Wheat	*T. aestivum*					
	Grass	*H. contortus*					

Contd...

Table 31–*Contd...*

Grasshopper Species		*Relative Host*	*Region*				
			a	*b*	*c*	*d*	*e*
Tribe–Trilotropidiini							
15. *Trilophidia annulata* (Th.)	Paddy	*O. sativa*	+	+	+	+	+
	Pea	*P. sativum*					
	Tomao	*L. esculentum*					
Subfamily–Hemiacridinae							
16. *Spathosternum prasiniferum prasiniferum* (walk.)	Lawn grass	*Cynodon dactylon* L	+	+	+	+	+
	Wheat	*T. aestivum*					
	Tomato	*L. esculentum*					
	Fenugreek	*T. foenum–graecum*					
Tribe–Heiroglyphini							
17. *Hieroglypus kolhapurensis* sp.nov.	Paddy	*O. sativa*	+	+	+	+	+
	Wheat	*T. aestivum*					
	Grass	*H. contortus*					
	Grass	*A. mutica* L.					
18. *Parahieroglypus bilineatus* Kirbay	Grass	*A. mutica* L.	+	+	+	+	+
	Grass	*H. contortus*					
	Grass	*T. quadrivalvis*					

Contd...

Table 31–*Contd...*

Grasshopper Species	*Relative Host*		*Region*				
			a	*b*	*c*	*d*	*e*
Subfamily–Oxynae							
Tribe–Oxyni							
19. *Oxya hyla hyla* (Serv.)	Paddy	*O. sativa*	+	+	+	+	+
	Tomato	*L. esculentum*					
	Grass	*T. quadrivalvis*					
20. *Oxya japonica japonica* (Th.)	Maize	*Z. mays*	+	+	–	+	+
	Paddy	*O. sativa*					
	Brinjal	*Solanum melongena* L.					
	Grass	*A. mutica*					
Subfamily–Cyrtacanthacridinae							
Tribe–Cyrtacanthacridini							
21. *Patanga succincta* (Johonsson)	Grass	*A. mutica*	+	+	+	+	+
	Grass	*H. contortus*					
	Grass	*T. quadrivalvis*					
22. *Cyrtacanthacris tatarica* (Linn.)	Groundut	*Arachis hypogaea* L.	+	+	+	+	+
	Soyaben	*Glycine max* (L.) Merr.					
	Tur	*C. cajan*					
	Jowar	*Sorghum bicolor* (L.)					

Contd...

Table 31–*Contd...*

Grasshopper Species		*Relative Host*	*Region*				
			a	*b*	*c*	*d*	*e*
23. *Anacridium flavescence* (Fab.)	Grass	*A. mutica*	+	–	–	+	–
	Grass	*H. contortus*					
	Grass	*T. quadrivalvis*					
	Paddy	*O. sativa*					
Subfamily–Romalinae							
24. *Teratodes monticollis* Gray	Ber	*Zizyphus jujuba* (L.)	+	+	+	+	+
	Badam	*Terminalia catappa* (L.)					
Subfamily–Catantopinae							
25. *Catantops pinguise* (Walk)	Paddy	*O. sativa*	+	+	+	+	+
	Grass	*T. quadrivalvis*					
	Lady finger	*Abelmoschus esculentus* L.					
26. *Xenocatantops humalis humails* (Serv.)	Paddy	*O. sativa*	+	+	+	+	+
	Lady finger	*A. esculentus*					
	Tomato	*L. esculentum*					
27. *Stenocatantops splendens* (Th.)	Ber	*Z. jujuba*	+	+	+	+	+
	Gliricidae	*Gliricidia sepium* (jacq.)					
	Paddy	*O. sativa*					
	Grass	*A. mutica*					
	Grass	*Dichanthium annulatum* (Forssk) Stapf.					

Contd...

Table 31–*Contd...*

Grasshopper Species	*Relative Host*		*Region*				
			a	*b*	*c*	*d*	*e*
Subfamily–Eprepocenemidiae							
28. *Choroedecus robustus* (Serville)	Sugarcane	*S. spontanum*	–	–	+	–	+
	Maize	*Zea mays* L.					
	Wheat	*T. aestivum* L					
29. *Tylotropidius varicornis* (walk)	Grass	*A. mutica*	+	+	+	+	+
	Grass	*T. quadrivalvis*					
Tribe–Eyprepocenemidini							
30. *Eyprepocenemis alacris alacris* (Serve.)	Brinjal	*S. melongena*	+	+	+	+	+
	Paddy	*O. sativa*					
	Jowar	*S. bicolor*					
	Pea	*P. sativa*					
	Green gram	*V. radiata*					
FAMILY–PYRGOMORPHIDE							
Subfamily–Orthacridinae							
31. *Colemania sphenarioidides* Bolivar	Grass	*A. mutica*	+	+	+	+	+
	Grass	*H. contortus*					
	Grass	*T. quadrivalvis*					

Contd...

Table 31–*Contd...*

Grasshopper Species		*Relative Host*	*Region*				
			a	*b*	*c*	*d*	*e*
Subfamily–Pyrgomorphinae							
Tribe–Chrotogonini							
32. *Chrotogonus oxypterus* Blanchard	Tur	*C. cajan*	+	+	+	+	+
	Green gram	*V. radiata*					
	Black gram	*Vigna mungo* (L.)					
	Grass	*H. contortus*					
Tribe–Atractomorphini							
33. *Atractomorpha crenulata* (Fabricius)	Wheat	*T. aestivum*	+	+	+	+	+
	Jowar	*S. bicolor*					
	Paddy	*O. sativa*					
	Grass	*A. mutica*					
Tribe–Tagstini							
34. *Tagasta apludi* sp. nov.	Grass	*A. mutica*	+	+	–	+	+
	Grass	*H. contortus*					
	Grass	*T. quadrivalvis*					
FAMILY–TETRIGIDAE							
Subfamily–Scelimeninae							
35. *Scelimena harpago* Serv	Algae	–	+	+	+	+	+
	Moss	–					

Contd...

Table 31–*Contd...*

Grasshopper Species		*Relative Host*	*Region*				
			a	*b*	*c*	*d*	*e*
	Water melon	*Citrullus lanatus* (Thunb.)					
	Cucumber	*Cucumi sativus* L.					
36. *Thoradonta purthi* Gunther	Paddy	*O. sativa*	+	+	+	+	+
	Wheat	*T. aestivum*					
	Cucumber	*C. sativus*					
37. *Acanthalobus cuneatus* Hanck.	Water melon	*C. lanatus*	+	+	+	+	+
	Cucumber	*C. sativus*					
	Tur	*C. cajan*					
Subfamily–Metrodorinae							
38. *Systolederus cinereus* Brunn.	Cucumber	*C. sativus*	+	–	–	+	+
	Water melon	*C. lanatus*					
	Wheat	*T. aestivum*					
Subfamily–Tetriginae							
39. *Tetrix bipunctata* (Linnaeus)	–	–	+	–	+	+	–
40. *Euparatettix personatus* Bol.	Wet land	*G. album*	+	+	+	+	+
	Weeds	*Stemodia viscosa* Boxb.					
41. *Hedotettix gracilis* De Haan.	Algae	–	+	+	+	+	+
	Water melon	*C. lanatus*					
	Tur	*C. cajan*					
	Wetland	*Gnaphalium album* L.					
	Weeds	*Bergia ammannioides* Roxb.ex Roth					

Contd...

Table 31–*Contd...*

Grasshopper Species		*Relative Host*	*Region*				
			a	*b*	*c*	*d*	*e*
42. *Hedotettix lineifera* Walk.	Algae	–	+	+	+	+	+
	Water melon	*C. lanatus*					
	Wetland	*Gnaphalium album* L					
	Weeds	*B. ammannioides*					
43. *Ergatetix dorsifer* (Walk)	Water melon	*C. lanatus*	+	+	+	+	+
	Cucumber	*C. sativus*					
	Wet land weeds	*B. ammannioides*					
44. *Ergatetix tarsalis* Kirbay	Algae	–	+	+	–	+	+
	Moss	–					
	Wet land	*S. viscosa*					
	Weeds	*Heliotropium ovalifolium* L.					
45. *Ergatetix guntheri* Stenmann	Algae	–	+	+	+	+	+
	Moss	–					
	Cucumber	*C. sativus*					
	Water melon	*C. lanatus*					
	Wetland weed	*B. ammannioides*					

Contd...

Table 31–*Contd...*

Grasshopper Species	*Relative Host*		*Region*				
			a	*b*	*c*	*d*	*e*
Family–Tettiigoniidae							
Subfamily–Conocephalinae							
46. *Conocephalus maculates*	Midges	*Contarina* sp.	+	+	+	+	+
	Small diptera	*Mephottix* sp.					
	Jassid	*Nilaparvata* sp.					
	Aphid	*Aphis* sp.					
Subfamily–Copiphorinae							
47. *Euconocephalus incertus* Walker	Grass	*A. mutica*	+	+	+	+	–
	Grass	*H. contortus*					
	Aphid	*Aphis* sp.					
	Jassid	*Nilaparvata* sp.					
Subfamily–Listroscelidinae							
48. *Hexacentrus unicolar* Serville	Cabbage butterfly	*Pieris* sp.	+	+	+	+	+
	Paddy borer	*Chilo partellus* Swinhoe					
	Jassid	*Nilaparvata* sp.					
	Aphid	*Aphis* sp.					
Subfamily–Mecopodinae							
49. *Mecopoda elongata* (Linnaeus)	Common Mormon	*P. polytes romulus*	+	+	+	+	+
	Mohagani moth	*Hyblaea machaeralis* Walker					
	Cabbage butter fly	*Pieris* sp.					
	Jassid	*Nilaparvata* sp.					

Contd...

Table 31–*Contd...*

Grasshopper Species	*Relative Host*		*Region*				
			a	*b*	*c*	*d*	*e*
Subfamily–Phnaeropterinae							
50. *Lentena intermedia* Ingrich	Grass	*A. mutica*	+	+	+	+	+
	Grass	*H. contortus*					
	Grass	*T. quadrivalvis*					
51. *Ducetia japonica* (Thunberg)	Wheat	*T. aestivum*	+	+	+	+	+
	Paddy	*O. sativa*					
	Grass	*T. quadrivalvis*					
	Cabbage butter fly	*Pieris* sp					
	Aphid	*Aphis* sp					
	Jassid	*Nilaparvata* sp.					
52. *Tapiena latifolia* Ingrisch & Shishodia	–	–	+	–	–	+	–
Subfamily–Oxylakina							
53. *Kirkaldyas truncate* (Karny)	–	–	+	–	–	+	–
Subfamily–Pseudophyllinae							
54. *Sathrophyla rugosa* (Linnaeus)	Ber	*Z. jujuba* (L.)	+	+	+	+	–
	Badam	*T. catappa* (L.)					

a: Kolhapur; b: Satara; c: Sangli; d: Ghats region; e: Plain region.

In all 30 species of family Acridiidae reported from Ghats and Plain region. Out of which *A. exaltata, G. africanus africanus, T. annulata, P. prasiniferum prasiniferum, O. hyla hyla* and *X. humalis humalis* were more abundant, they survive in all climatic conditions of the year and found feeding on variety of plants and breed throughout the year.

In the Present study eleven species of family Tetrigidae reported from Ghats region *viz. S. harpago, H. lineifera, T. purthi, A. cuneatus, E. personatus, E. dorsifer, E. tarsalis, E. guntheri, T. bipunctata, H. linefra, S. cinereus*. Out of which 10 species were also reported from Ghats region. *T. bipunctata* is not reported from Plain region. All are reported from August to February. In family Tettiigoniidae 9 species have been reported from Ghats region and 6 species reported from Plain region. Out of which, *D. japonica, H. unicolar* and *L. intermedia* were found in July to August. Maximum population was reported in the September and minimum population reported in July while, *E. incertus* and *M. elongate* were abundantly reported in June to November.

G. semipictus, T. monticollis, P. bilineatus and *C. robusta* species of Family – Acrididae and *S. cinereus* species of family Tetrigidae were rarely reported from Plain region.

I. convergence and *A. flavescence* from family Acrididae, *T. bipunctata* from family Tetrigidae and *K. truncate, T. latifolia* and *S. rugosa* species of family Tettiigoniidae rarely reported from Ghats region and all these species are lacking in plain region and *G. semipictus* and *C. robustus* species lacking in Ghats region.

Paulraj *et al.* (2009) studied distribution of different host plants and habitats in nine localities of northeastern Tamil Nadu, for grasshopper species, from August 2004 to 2006. 33 species of grasshoppers were reported belonging to four families. Family Acrididae was found to be the predominant group of grasshoppers represented by 21 species. Family Tettiigoniidae was the second largest group represented by six species failing under five genera and three subfamilies. 18 species of grasshoppers were found associated with grasses and 11 species found as a surface grasshoppers. The species *Poekilocerus pictus* (Fab.) was reported with high distribution.

In the present study the distribution of 54 species of grasshopper from Satara, Sangli and Kolhapur districts have been reported. Out of which

P. pantali, I. convergence, D. venusta, C. robustus from family Acrididae; *S. cinereus, E. tarsalis* from family Tetrigidae and *T. latifolia* and *K. truncate* from Tettiigoniidae recorded the less distribution from study area.

Shishodia *et al.* (2002) published checklist of Orthoptera from Pong dam Wetland, Kangara District, Himachal Pradesh, India, He reported 39 species belonging to 35 genera, under eight families. Thakur *et al.* (2004) prepared checklist of Orthoptera from Roper Wetland Panjab, and recorded 42 species of Orthoptera belonging to 12 families and 38 genera. A checklist of Orthoptera from Pin valley national park, Lahul and Spiti, India was made by Thakur and Mattu (2006). They recorded 10 species of Orthoptera belonging to family Acrididae under four subfamilies.

Chandra *et al.* (2007) prepared the Checklist of Orthoptera of Madhypradesh and Chattisgarh. They reported 139 species of grasshopper belonging to 12 families. Shenthilkumar *et al.* (2006) made a checklist of Orthoptera from Giboon Wild life sanctuary Assam and recorded 25 species of Orthoptera belonging to 21 genera and 12 subfamilies while, checklist of Orthoptera prepared from Kolhapur,Satara and Sangali districts contain 56 species of Grasshoppers belonging to four Families, 20 subfamilies, and 49 Genera. Recently, Bhusnar and Sathe (2011) studied the diversity of grasshoppers from Kolhapur districts including Western Ghats of Maharashtra. They reported 19 species of family Acrididae, 2 of Pyrgomorphidaes, 9 of Tetrigidae and 9 of Tettiigoniidae, found feeding on agricultural and forest plants. In general the diversity of grasshopper was quite rich in agroecosystems than that of forest ecosystem of Maharashtra. Similarly, Bhusnar and Sathe (2012) studied ecology of grasshoppers from Kolhapur district Maharashtra. They reported 40 species of grasshopper belonging to four families it was found that family Acrididae was diversified well adopted fore different host plants and found throughout the year. Minimum population of the family Pyrgomorphidae was recorded. The population of family Tetrigidae and Tettiigoniidae found were only monsoon and post monsoon period. However, maximum population of grasshopper was reported in the month of September and minimum in the month of April from Kolhapur region. The present work will be helpful for checking the population of grasshoppers and the data will be helpful to ecologist and environmentalist.

Chapter 6
Biology of Grasshoppers

The order Orthoptera contain 20,000 species. The grasshoppers and locusts are visualized as most destructive pest insects of this order. Some grasshoppers cause significant damage to tree seedlings and agricultural crops. They are oligophagous and classified as graminivorous, forbivorous and ambivorous. Being important component of food chain of birds and mammals they have special importance in ecology. They have three stages in their life, *viz*. egg, nymph and adult. They found in copulating several individuals with opposite sex. Thus, they are promiscuous. The mature male abruptly mounts the female however, copulation takes place only if the female is in a receptive stage.

The time taken from the last moulting (ecolosion) to reach sexual maturity is called the pre-copulation period. The pre-copulation period varies with the species. In *Atractomorpha crenulata* it is 4 days for males and 13 days for females after last moult; in *S.P. prasiniferum* males copulate with the females at an average of 3.5 days, after emergence while, *Parahieroglyphus bilinealus* copulates 6 to 8 days as males and within 10 to 15 days as females.

According to Agarwal (1955) the pre-copulation period is also season specific and occurrence too. In *A. crenulata* it was 10-13 days in September, 9 days in March, 7 days in April, 5 days in May, 4 days in July, 5 days in August and 9 days in October.

According to Katiyar (1960) there are three modes of copulation found in short horned grasshoppers. In mode (I) the male rides on the back of the female head to head. In mode (II) the male positions on the side of the female and in the (III) mode, the male neither rides on back of the female nor lies at her side, but hanges positively, attached to the abdomen during coitus. An additional intermediate "dorso-lateral" mode was also observed in *Phlaeoba panteli*.

Riding was the normal mode of copulation (Mode-1). In this method the excited males clasp the female from the back. Sometimes two or three males competed for riding on the back of a female for coitus. However, only one male succeeded in riding. Head to head and the terminal ends of the two partners abdomens inter-locked with each other. Then the male abdomen was curved down to either right or left of the female.

The copulating partners go on feeding during copulation in some cases and in other cases the male usually remains on the back of the female even after coitus was over and later in the act of oviposition also.

In the field in *Phlaeoba antennata* at Jalpaiguri, *Cyrtacanthacris tatarica* at Chilka Lake and *Paraconophyma scarab* at Simla Hills, it was observed that the copulating pairs were completely engrossed in the act; the insects were motionless and unconcerned about the surroundings even when they were disturbed. They stayed on the palm for 55, 40 and 35 minutes respectively. The pair of *P. scabra* was interlocked for very longer period and were unable to release themselves. The duration of copulation varied from 19 minutes to 5.48 hours in *A. crenulata* and in *Acrida exaltata* from 3 to 14 hours (Bhowmik and Halder, M.S.) while in *S.P. prasiniferum* it was 8 to 49 hours (Iqbal and Aziz, 1974). However, Pradhan and Peswani (1962) recorded in 4-42 hrs as interlocked period in *Hieroglyphus nigrorepletus*.

The season of copulation and oviposition are closely interrelated in grasshoppers.

The process of egg laying on/in the substrata is called oviposition. The successful copulation initiates oviposition process. Oviposition does not occur immediately. Hence, the time taken from copulation to oviposition is called pre-oviposition or post- copulatory period. The pre-oviposition period is variable from species to species and also in the same species depending on season. In *A. crenulata,* it was 15 to 16 days in September,

13 to 14 days in March, 11 to 12 days in April, 10 to 11 days in May, 9 days in July, 6 days in August and 10 days in October while in *P. bilineatus* females it was 9 to 12 days.

In general, the pre-winter copulating period in females was different in different species. Usually in species having one breeding season copulation, oviposition and hatching take place round the year and a population of both adults and nymphs were found during severe winter and in rainy season of the same year.

The period of incubation of eggs is also dependant on seasons. *Aiolopus thalassinas* oviposites late in October, January to March in Transval, whereas in cape peninsula in April to May (Chester, 1938).

Oviposition lasted for 3-4 hours or slightly more in *Acrida etaltata;* however, in few cases it required half an hour to one hour (Bhowmik and Halder, N.S.). In *P. bilineatus,* 2 to 3.5 hours are taken although in a few cases it may be only 1.5 hours. In *S.P. prasineterum* the duration was 2 to 4 hours, while in *H. nigroreptelus* it was 2.30 hours. In species where the females oviposite more than once, the duration of oviposition decreases with subsequent ovipositions. After 3 to 16 days of copulation the female look 'lazy' and start making holes in the soil with the help of the ovipositor.

The final hole for egg laying is made by making a few trials. According to Bhowmik and Halder if the breeding ground was unsuitable females of *A. exaltata* tried more than twenty times for suitable place.

Oviposition takes place in moist and dry soil. During oviposition female extend its abdomen slowly in soil and it is almost doubled to its normal length. Most of females of grasshopper species oviposit by this way. While ovipositing female make an angle of 10° to 65° with the body above the surface of the ground.

The female lay eggs in egg purse / egg pod. Oviposition and the orientation in the egg pod almost same, in many species; However, in some there are fine differences. The female release a frothy substance at the bottom of the hole and then lay the eggs one by one after a little pause between two successive eggs; The tips of the dorsal valvulae of the oviposition help in filling the eggs in egg pod. After filling all the eggs, the female plugs the egg pod either by making a fragile, frothy cap about the eggs (in most cases) or with a hard cap above the eggs with some cement material; deposited above the cap. After release of eggs, female withdraws the abdomen from the hole and scraps the surface around the hole with its posterior legs as part of parental care.

Female grasshopper lay eggs in egg pods at a depth ranging 2.5 to 12.5 cm and 3.2 to 7.21 cms below sandy loam. The depth of egg laying vary with length of abdomen and texture and moisture of the soil.

The famels generally avoid to lay eggs in the middle of the fields, in some cases in gardens and nurseries etc. However, in most cases oviposition occurred during September to October when the fields were either waterlogged, innundated or flooded. The females choose sites along the sides of frodden paths, embarkments of agricultural fields or nearby high lands, along cart-routes, bank of rivers and along dense bushes of foot hills in high grasslands.

The grasshoppers exhibit strong preference for various soil conditions, may prefers xerophilic (dry) soil for oviposition, others prefer mesophilic soil without abundance of moisture although top layer being moist) and some to hygrophilic (moist) soil. If a gravid female do not find out a suitable site, it either drops its egg pod on the soil surface, grass leaves or on glass jar wall (eg. *Acrida).* In extreme cases stops taking food and ultimately dies.

In an aquatic acridia, in extraordinary circumstances, as exhibited by the females *Gesonula punctifrons* (Stal.) egg-laying takes place on the succulent stem of the Colocasia (Semba) plants growing along the field bunds. Around black hole of about 4 mm in diameter is made on the stem. However, the females of the genus *Oxya* oviposit usually in drill soil and in flooder fields, to cement egg-masses between grass stems, in ley axils or on stones and one or two inches above the water level (Uvarou, 1928). According to Iqbal and Aziz (1974) *S.P. prasiniferum* lay eggs in floating over still water.

According to Katiyar (1956) the female grasshoppers are usually prolitic breeders. A female may lay eggs once (eg. *Acridia*), twice, thrice (*A. punctatus*) or even upto 5 times (*e.g. P. bilineatus*) in its life time. The number of eggs in each egg pod varies from female to female, depending on climatic factors supply of food and geographical variation. In *Hieroglyphus banian*, in the insectory, gives about egg-pods per female, and the number in each pod varies from 29-49 while in Assam each female of the same species lays 50-60 eggs altogether (Chowdhury and Majid 1954) and 68-90 eggs (average 81) in Uttar Pradesh (Gupta and Saxena, 1963). In *Melanoplus bilituratus* in the Western Canada, higher fecundity was recorded where hatching took place early in the season. Grasshoppers in one case of the earlier-hatched group produced 22.7 egg pods per female. This high fecundity was partly due to the greater adult longevity and to a high rate of oviposition. The same author further mentioned that the rate of oviposition, in terms of egg production per female per day, was highest about mid summer (3.4 to 5.3 eggs per female per day) when temperature was at its maximum. The decline to the rate of oviposition (1.7 to 2.6 eggs per female per day) towards the end of the season was attributed mainly to declining temperatures and not to the old age of the females.

In *M. bilituratus* Parker (1930) recorded 6 to 8 egg pods per female when reared at temperatures ranging from 27°C to 37°C, but only 0.4 egg-pods per female was noted at 22°C.

Fecundity is affected by different food plants (Pickford, 1958, 1962) climate (Pick ford, 1966) temperature and moisture (Khan, 1974; Me carthy, 1956). The above factors have immense effects on fecundity rate and fostering readily available of luxuriant growth of vegetation for the newly hatched nymphs, and ultimately check grasshopper population.

In *Gesonula punctifrons* minimum egg production 8 to 12 was recorded. In *Acrida exaltata* only one egg-pod was provided in its life with 47-66 eggs. The female of *A. crenulata* laid 3 to 5 egg pods, each egg-pod contained minimum 64 eggs (average 21.3 per pod) in September and maximum 136 eggs (ave. 27.5 per pod) in April. In *Schistocereus gregaria* the average number of eggs per pod was found to be 54 in Rajasthan desert (Singh and Singh, 1978). The female of *S.P. Prasiniferum* showed 6.40 ± 0.393 egg pods with average 51 ± 2.704 eggs. In *Hieroglyphus concolor* each egg pod showed 62-84 eggs (Katiyar, 1960) to 123 eggs (Gupta and Saxena, 1963).

The duration of egg hatchability has great significance in completion of life cycles in a single year. Short life cycle leads to increase the population of species. The incubation period of grasshoppers should be increased as much as possible for suppression of their populations. However, shortest period of incubation (10 days) in *S. gregaria* was recorded by Singh and

Singh (1978) in natural conditions. *Gastrimargus africunas* showed 2 or more generations in a year. The incubation period was 22-24 days (Deseamps, 1965) likely in *S.P. prasiniferum* it was 20-22 days while, in *A. exaltata,* with 2 annal cycles the incubation period was 23-27 days. In species having one annual cycle the period was moderately long while in biannual and multiple year life cycle it was considerable long. A grasshopper *Arphia conspersa* scudder, having a two year life cycle, over winters as a partially grown up nymphs or as an egg in the first winter and a partially grown nymph in the second winter (Pickford, 1953). In *Melanoplus alpinus* scudder, a species of high elevations, having a multiple year life cycle (requiring 3 years) was noticed by Kreasky (1960). At the time of hatching, the egg-chorion becomes dry and splits longitudinally. As development of eggs depend on the moist condition of egg pod environment. In the insectory conditions all egg pods should be kept moistened by applying water.

The egg-pod is like a bag or capsule and almost straight and cylindrical tube which has a heart or concave anterior ends with depressed or concave area known as "cap". The posterior end is rounded or tapering.

The egg-pod is made up of frothy and spongy substance secreted by female during oviposition which hardens on exposure to air. The structure of pod is species-specific and is dependent on nature of soil.

Egg-pods deposited in the xerophilic (dry) soil are simple and short but their outer wall is very rough, tough, hard and cemented. The concave caps are hard, smooth and placed either just above the layers of eggs or at the apex of the pod. In the mesophilic pods the outer wall is not cemented, but is rough, slightly hard and consist of a few layers of dense, laminated, secretions. It may be with or without a cap at the top. The cap, if present, is usually soft and membranous. In the hygrophilic (with abundant water) pods, the outer wall is smooth or rough, spongy or troty, but tragine, it is without a cap but the top section is filled with coarsely meshed, spongy secreted material with septra or hexagonal empty cells.

Variations are noted in pods in their external shape. They are curved with one side concave and other side convex. In *Phlaeobs panteli, Choroedocus insignis* and *P. bitineatas* they are broad in the middle and apical end is conical or posterior end is broadened and rounded or constricted at apical end. In grasshopper, size of pods is highly variable. It may be twice as long as wide (length 1.6-1.7 cms; width 0.8-1.0cms) in *Hieroglyphus assamacris* or about 4 times as long as wide (length 1.45-1.85cms; width 0.45-0.6cms) in *Cercaris deplorata* or 8 times as long as wide (length 2.7-3.8cms; width 0.35-0.5cms in *Eypvepocnemis rosea.*

Eggs are tiny (egg size was 3.25- 4.15 X 0.85-0.95mm) cylindrical or subcylindrical whitish bodies. In *O. abruptus* apical end or anterior pole is rounded. At posterior micropylar rings are distinctly visible, being darker than the rest of the egg surface. The egg wall or chorion in the micropylar region shows a series of minute micropylar canals, each having an external and internal openings and roofs. The structure of egg is studied by Katiyar (1960) which has great important in the systematic studies of Acrididae. According to Katiyar (1960) the Sculpturing and structure of egg chorion is species specific.

The arrangement of eggs in egg- pod is also species specific. They may be arranged vertically at right angles or obliquely one above the other and in many other ways but always with the micropylar poles pointing towards the base of the egg- pod. Difference in egg size, shape and colorations are also well illustrated in grasshoppers. Larger eggs may be up to 8 times bigger than the smaller ones. The eggs of AK grasshopper *Poekilocerus pictus* Fabricius (Acridiidae) are orange coloured and about 145 to 170 elongate eggs are laid in a spiral manner to form a compact mass in a egg pod. The eggs laid in summer over winter for about 4 months while in Rice grasshopper *Hieroglyphus banian* Fab. egg laying is continued from September to November. In each egg pod 30-40 eggs are laid.

The freshly hatched nymphs are "vermiform". Such vermiform larvae are enclosed in a thick membrane with their head bend towards the antennae and legs lying closed to the body.However, larvae free themselves from the unsheathing membrane by reaching the soil surface. This worm like larva is very tiny measuring about 2.00 – 3.5 mm. after some time this larva is converted into the first stage nymph or instar.

The nymphs of many short- horned grasshoppers undergoe 4 to 7 moults before reaching the adult stage. In *Gesonula punctifrons* 4 moults are noticed within 22 days, at an interval of about a week (Ayyar and Menon,1933). There are about 5 to 6 stages in *G. ayricanus* which lasts for

30-39 days. In *Hieroglyphus nigroptelus* the nymphal period was 3 weeks (Roonwal, 1945) while, Grist and Lever (1969) reported as 71 days at 26°C and 35°C (average 32.5°C ± 1 2.5°C) nymphal period in the same species. The number of moults and total period taken to complete its life cycle varies from species to species. In female generally one stage is more than the males the duration of each moult also varies to some extent depending on season and availablility of food. The nymph stops feeding and become sluggish before moulting.

Newly hatched nymph can moult within 1 to 2 days. The duration of first moult may be varying with the species. The nymphal period is generally longest in prewinter season (November, December and sometimes in January) and shortest during full mansoon (July- August). In *A. crenulata* the nymphal period was longest (31-69 days) during December and January and shortest during July and August (31- 39 days). Hatchability of eggs into in nymphs and nymphs to adult is 100 per cent. However, grasshoppers have high fecundity rate.

The nymphs are characterized by having incomplete development of antennae, tegmina and genitalia. Nymphs more or less resemble with the adults but lacking wings and are smaller in size and different in coloration. The coloration often differs markedly from adults. Sexes are not recognizable before 2nd stage and clearly distinguishable during 3rd stage. According to Roonwal (1946) in 5th stage, the nymphs usually have the tegmina somewhat developed but with the anal area up. In the successive stage, the position of tegmina rotates up to 90°C. This directional rotation of tegminal axis brings costal area up and anal area down. This down position of costal portion of tegmina at 5th stage differentiates the nymph from a brachypterous (Semi – apterous) species.

Nymphal duration and number of instars are varying with the species. In *H. banian* there are 6 instars and nymphal perioed is 3 weeks while in AK grasshopper the nymphal perioed is 30 to 43 days. The nymphs of this grasshopper are yellowish with orange and black stripes and dots all over the body.

Short horned grasshoppers have 1 to 3 annual breeding seasons. However, large species shows only one breeding cycle. In *H. banian*, *H. nigrorepletus* and *P. bilianitus* only one breeding cycle was noticed with greater interval between breeding season. Phipps (1970) Says that the number of cycle (5) depends chiefly on the size (Weight of the gravid female). Out of 17 small species (Weight under 500gms), he recorded two species with one cycle, eight with 2 cycle and 7 showed continuous breeding. However, 3 medium sized species (500- 1000gms) were with one cycle, only one large species, *Eyprepocenemis plorans,* showed continuous breeding with species having one cycle. In most cases, egg laying period extended from the beginning to the end of the rainy season and in other species, it started at the beginning of the next dry season. While in small species oviposition begins a few weeks before the onset of the next rainy season.

In larger species eggs are laid early to counteract the dry as well as the wet season. Large eggs may be up to 8 times the volume of smaller ones. According to Chapman (1962) the development of large eggs requires more absorption of water than smaller eggs before development begins. This requirement of water may account for the restriction to one breeding season by large species. In a year with having one rainy season in the place of its occurrence. Change in the breeding behaviour of a species was noticed with respect to the geographical pattern of the environment. In *Gastrimargus africanus* (Sauss.) there were normally 2 cycles in arid zones (Descamps, 1953), but there were 3 in Madagascar (Descamps and Winterbert; 1956) and was continuous breeder in Ghana (Chapman, 1962). Size (Wt. of ovaries) has no relation with as pointed out by Pickfored (1960) with exact stage of breeding cycle of the hopper; Overlaping of least 2 annual cycles are noticed in *Oedelus abruptus* wherein eggs overwinter from November to April (Ahemed *et al.,* 1973). In *O. senergalensis* the size was almost same previous one, in 3 cycles (Launois,1979). In African small species, *Spathosternum prasiniferum prasiniferum,* breeding found throughout the year (Iqbal and Aziz, 1974).

All the eggs laid are neither hatched as nymphs nor all the nymphs reach to adult stages. The rate of fertility and survival rate was intensely proportional to mortality and other ecological factors. The rate of mortality chiefly was dependant on climatic conditionas.

Singh and Singh (1978) worked with *S. gregaria* in Rajasthan desert, under natural field conditions at minimum 73.6°F to maximum 106.3°F temperature and relative humidity 49 per cent -89 per cent and rainfall 0.22 – 0.85 inches found that there was only 9-10 per cent mortality in egg stage, whereas in the laboratory it ranged between 10 per cent–60 per cent (Husain and Ahmed, 1936) under different conditions of temperature and humidity etc (at 30°C with 100 per cent humidity, 40-90 per cent hoppers hatched out). In *S. gregaria* with a constant temperature of 20°C–21°C, 75- 83 per cent of eggs hatched. In *S. prasiniferum* the fertily rate of eggs were 91.76± 1.06 per cent (*i.e.*, about 8-9 per cent mortality) at constant temperature of 33°C ± 1 °C and 70 per cent ± 5 per cent pH, with 12hr alteration of light and darkness. According to Agrwal (1955) 53.6- 66.6 per cent of nymphs of *A. crenulata* can reach to maturity.

The longevity of grasshopper has direct relation with the increase population and increased damage to crops. At insectory condition, *A. exaltata* lived for 34 and 64 days. However, two females of above species lived as long as 129 days. Male of *P. bilineatus* lived for 32-43 days and female for 58–64 days, the longevity of *A. crenulata* was 25 – 36 days and 30-68 days for male and females respectively. However, the male of *S. prasiniferum* survived more than females (29 to 40 days) (Iqbal and Aziz, 1974).

After copulation males normally die but can survive for few days. However, females may die either immediately after oviposition (*A. miliaris* Linn.) or later (*A. exaltata*) (Bhowmik and Halder). The population may have relation with the seasons. *M. bilituratus* hatched in May showed increased population up to times than those hatched 7 to 9 weeks later in July. However, the most vulnerable stage in the whole life cycle appeared to be the very early nymphal stage. The mortality rate was lower in older instars than younger instars. It was observed that adults were able to withstand heavy rain or extended period of cool weather.

Chapter 7

How to Control Grasshoppers

There are several methods of control of grasshoppers. Important methods of grasshoppers are given below.

1. Preventive control
2. Ecological control
3. Biological control
4. Cultural control
5. Microbiol control and
6. Chemical control

1. Preventive Control

i) Grasshopper nymphs and adults are collected and dipped into kerosinized water or pesticides, emulsifiable concentrates for killing them.

ii) Grasshoppers lay eggs in soil in egg pods. Therefore, deep ploughing or digging the field can expose the eggs to natural mortality factors like biotic and abiotic and kill the eggs.

iii) Collection of grasshoppers may be done with net, tins coated with some adhesive like coaltar,

iv) Clean cultivation and removal of grasses.

2. Ecological control

Ecological control has two important components *viz.*

a) Abiotic factors

b) Biotic factors

a) Abiotic Factors

Abiotic factors such as topography, water bodies, soil quality, climate, temperature, humidity, light, rainfall, etc. play important role in control of grasshoppers.

Temperature, rainfall and snowfall are largely responsible for out breaks of grasshoppers. Temperature is the most important factor which determine the size of the spring grasshopper population. Warm days, spring and summer determines population dynamics of grasshoppers including parent and current grasshoppers. Climatic conditions in the fall are the limiting factor of successful egg laying and thus influence the number of eggs laid. Temperature also affect embryonic development, thereby affecting the time of hatching. In spring, development is slowed down in grasshopper or stopped.

The effect of cold winter temperatures on grasshopper egg survival is minimal. The eggs can survive at 15°C in soil, wind current affect egg survival. Similarly, snow cover and temperature 40°C extend the hatchability of eggs in number of days.

Spring temperatures have minimal effect on the survival of the newly hatched grasshoppers. Young grasshoppers are hardy enough to survive low, even below freezing spring temperatures, providing these temperatures do not persist for several days. The spring temperature can affect the grasshopper development and plant growth. In cold, development is slow and in hot it is faster. If the spring is hot, grasshoppers will hatch early and develop quickly. Cool spring temperature will lead slow development. Crop development is also affected by less than ideal temperatures and thus, food requirement of the grasshoppers.

It is observed that relationship between temperature and rainfall controls the amount of crop damaged by grasshoppers. Under hot dry conditions, a small grasshopper population can cause much damage as a large grasshopper population under cool and wet conditions.

Environmental moisture can affect the size of the grasshopper population. During an extended drought, lack of water may slow the development of many eggs and can destroy eggs, especially during certain embryonic stages including eclosion. However, extremely dry conditions can kill grasshopper embryos under drought conditions.

Rainfall has less effect on grasshopper populations. Rainfall will only have an effect with heavy downpour, may reduce population immediately after an extensive hatch. However, a cool, wet june will not seriously affect grasshopper populations.

The cool and wet weather reduce the crop losses by hindering grasshopper's development and increase the diseases in the grasshopper population and thus, helping to reduce next years grasshopper population.

River side grassland help grasshoppers to maintain their population but large water bodies are not helpful in this regard.

3. Biological Control

Natural enemies also play an important role in controlling grasshopper populations. In some localized areas these enemies may even be more important factor than the weather. Some of the grasshopper enemies attack eggs in the soil while others attack nymphs. The eggs of grasshoppers are eaten by blister beetle grubs in soil. Thus, the grub act as good biocontrol agent of grasshoppers.

Egg Predators

The eggs of grasshoppers are predated by predators like bee flies, blister beetles, ground beetles, crickets and other insects.

The adults of some of these insects, like the common field cricket, feed directly on the eggs and may destroy upto 50 per cent of the eggs in some areas. Other egg predators like bee flies and blister beetles, deposit their eggs in the soil near grasshopper eggs.

When the larvae of egg predators hatch, these larvae locate the egg pods and feed directly upon the grasshopper eggs. The bee flies and blister beetles can destroy upto 70 per cent of the eggs in soil.

Egg Parasites

The wasp belongs to the genus *Scelio* deposit their eggs within the newly laid grasshopper eggs. The parasitic larvae complete their development within the eggs of grasshoppers and emerge as an adult instead of young grasshopper. The newly emerged parasitic flies again parasitizes the egg of the next generation of grasshoppers.

Nymphal and Adult Predators

Spiders, robber flies, wasps and many birds are recorded as predators of grasshoppers. All above predators consume them in large numbers. Thus, affect the population of grasshoppers on large extent.

Nymphal and Adult Parasitoids

Parasitoids contain flesh flies, robber flies, muscoid flies, tangled vein flies and many others. The above parasitoids lay their eggs either on nymphs or adults. After hatching the eggs parasitoid start eating internal tissues of the grasshopper and kill them affecting the population of grasshoppers.

Microbiol Control

The suppression of pest population below the level of economic damage by using pathogens is called microbial control of pest. Following microbs are used for the control of grasshopper populations.

(i) Nematodes
(ii) Protozoans
(iii) Bacteria
(iv) Fungi

i) Nematodes

Threadworms are frequently found coiled inside grasshoppers. The "threadworms" are remain overwinter in soil and lay their eggs in the soil or on vegetation. Eggs or juveniles are treated on crop for contamination of food of grasshoppers. When treated crops are eaten by the grasshopper, the worms will enter into the alimentary canal of grasshopper and multiply in the body of grasshopper and finally kill the hoppers. In nature they are available in soil and get entered into the body of grasshoppers through ingestion. Later, multiply in the body and kill the grasshoppers.

ii) Fungi

Entomophaga grylii is effective controlling agent of grasshoppers under, humid conditions. This fungus may occasionally reach epidemic proportions. The disease leaves the corpses of its victims, cliging to the stems of plants. The fungi release toxins into the body of grasshopper which affect protein synthesis and kill the grasshoppers.

iii) Protozoans

The naturally occurring microsporidian parasite *Nosema locustae* has been found effective biocontrol agent of grasshopper populations. The systic and non systic forms are found attacking grasshoppers. If such forms ingested by grasshoppers, they multiply in the body and kill the grasshopper. The protozoan disease is contaminative. Hence, cause higher mortalities in grasshoppers. The protozoan can reduce a grasshopper population upto 5 to 40 per cent in one year. The parasite can reduce feeding rate to as one-third of normal. Reduction in the number of eggs was also noticed in protozoan infected pest grasshoppers. However, this organism as a biological control agent have shown limited success.

There are good number of natural enemies for grasshoppers and those enemies could be used to prevent grasshopper outbreaks over extensive areas. Therefore, there is need to develop mass production technique for natural enemies so that those can be used on large extent for control of grasshoppers.

Nevertheless, natural enemies do play an important role in controlling localized grasshopper populations and keeping environment ecofriendly.

4. Cultural Control

The cultural control of grasshopper is least expensive. Under this method following strategies are visualized.

i) Early sowing
ii) Crop rotation
iii) Tillage
iv) Trap trips/crops
v) Manipulation of Environment

Manipulation of Environment

The proper timing of operations can play important role in the production of a crop. By modifying the grasshoppers environment at certain critical periods of its life cycle, one can reduce grasshopper numbers directly or can, affect ability of grasshoppers to reproduce.

The cultural control methods should be used reluctantly because it is difficult to assess their effectiveness. Nevertheless, these methods have great relevance in keeping environment ecofriendly. Cultural methods take time to work but it is best preventive approach of pest management including grasshoppers.

Early Sowing

Crops should be sown as early as possible. Older plants can withstand with more grasshopper feeding than younger plants. Early seeding will not prevent crop damage entirely but reduce the amount of damage to crops and will provide also more time for farmers to keep crop insecticide free. Early seeded crops mature early and grasshoppers are less attracted to them as they are to lush young foliage.

Crop Rotation

Whenever possible, avoid sowing cereals on stubble fields heavily infested with grasshoppers. Cereals should be sown on stubble fields only where soil moisture is adequate and where one or more applications of an insecticide are given.

Tillage

Cultivation of the soil is very good cultural practice performed by farmers for the reduction of grasshopper populations. Very carefully, especially under drought conditions. Tillage controls grasshoppers primary by eliminating the green plants on which grasshoppers feed. Hence, this method has little value. Excessive tillage is harmful; it will reduce soil moisture level and increase the risk of soil erosion.

Weeds encourage grasshoppers to lay eggs upon them and early fall will discourage female grasshoppers from depositing their eggs in the fields. Grasshoppers lay eggs in clean summer.

It is advisable to complete early spring tillage or give the chemical treatment to eliminate all green growth on stubble fields before the grasshoppers have hatched and deprive the availability of food for young grasshoppers. Grasshoppers will die due to starvation. Early tillage provides additional benefits that it gives good weed control and conserve moisture at no extra cost.

Tillage can be used as a last choice "hot spots" where young grasshoppers are continuing to hatch in large numbers and when chemical control is not desirable. Burrowing the eggs and hatching grasshoppers deep enough can avoid the emergence of young hoppers to the surface.

Trap Strips

Adequate control is possible by simply climinating all green plant materials in a field. Once grasshoppers have fed and developed to the second stage of growth in a field, they are usually move to adjacent crops when their existing food supply is exhausted. A trap strip is made by cultivation of black guard strip of lom wide around the outside of a field. Additional trap sites may be made as per the intensity of grasshoppers. The black guard strip should be sufficient for grasshoppers to move promptly into the trap strips to feed. Trap strip should contain less host vegetation to feed for only 1-2 days for grasshoppers.

The strips of wheat or spring rye are more effective with seeding several weeks before tillage begins. The young grasshoppers take several days for migration from the cultivated guard strips to the trap strips. After complete migration the trap strips should be treated with an insecticide since young grasshoppers are very susceptible to insecticides. The highest recommended doses of insecticide be used for adequate control of grasshoppers. 3 days frequency be kept for treating crops for adequate control of grasshoppers.

Economic Thresholds

The economic threshold depends on several factors such as :

i) Stage of the insect
ii) Type of crop
iii) Crop stage
iv) Growing conditions
v) Cost of control and
vi) Current market value of the crop.

The economic threshold of grasshoppers have been detected in cereal crops. The most serious economic damage is detected with third to fifth nymphal stages. The economic threshold in cereal crops ranges from 8 to 12 grasshoppers per square meter.

Oat crop is less preferred as food by the grasshoppers. Similarly, peas are also less preferred. In both these cases, even if grasshoppers do feed on the crop, damage is very limited. These crops affect reproductive potential of grasshoppers negatively. Therefore, these crops can be used as a trap crops.

Lentil is more susceptible to grasshopper feeding than other crops. Grasshoppers feed on Lentil pods, flowers and leaves including early minute pods. The feeding on the pod is the risk of diseases and staining of the seeds which affect yield and marketability of the product. Two grasshoppers per square meter can affect the yield adversely. The crop canola is not preferred by grasshoppers but at very young stage and again when the pods are ripening damage intensity of grasshoppers is increased. However, according to some workers significant damage can be done at all stages of growth when grasshopper pressure is high.

The migratory grasshoppers damage pods primarily by chewing holes into the pods. The two-stripped grasshoppers generally remove whole strips from the sides of pods. Both grasshopper species cause considerable damage to crops. Hence, insecticide application is needed for their control.

In India short horned grasshoppers cause severe damage to cereals like Jowar, maize, paddy, Bajra, etc. They are also destructive to sugarcane, leafy vegetables, certain pulses, fruit crops, groundnut and several other agricultural crops. Therefore, they should be controlled by above said methods including chemical methods.

Chapter 8
Conclusion

In present study, 54 species were reported belonging to four families, 21 subfamilies, and 47 genera, from Satara, Sangli and Kolhapur districts of Western Maharashtra, three new species namely *Stauradorus indica* sp. nov., *Pternoschirta sahyadricus* sp. nov., and *Tugsta apludi* sp. nov. have been described for the first time. Out of 54 species, 10 species were rare and 44 were common. The Common species were responsible for extensive damage to plants by defoliation and for decreasing the growth and yield of agricultural and forest crops.

Out of four families, family Acrididae was more dominant with 30 species over the Pyrgomorphidae 4 species, Tetrigidae 11 species and Tettiigoniidae 9 species. The members of family Acrididae and Pyrgomorphidae were abundantly scattered in agro, forest and grassland ecosystems while, family Tetrigidae members reported as abundant during winter in Wetland ecosystems and members of family Tettiigonidae were abundant in forest and agro ecosystems. Some members of family Tettiigoniidae namely *C. maculatus, H. unicolar, M. elongate, L. intermedia* and *D. japonica* were found to be very good biocontrol agents of agricultural crop pests in the Western Maharashtra.

Although Western Ghats is among the 18 biodiversity hot spots of the world. Very little attention is paid on the protection and conservation of biodiversity. The objectives of biodiversity conservation and protection can be achieved by more and more involvement of scientists, teachers, students, farmers, common people, women and children too.

Chapter 9
Summary

Western Ghats is one of the rich center for floral and faunal diversity and among 18 biodiversity hot spots of the world. It is the habitat of many endemic rare and endangered species of animals and plants. However, very little information is available about insect biodiversity of Western Ghats. Most of the work related to insect biodiversity is carried out from southern region of Western Ghats. Hence, the present work was carried out from Western Ghats of Maharashtra particularly Kolhapur, Sangli and Satara districts.

Grasshoppers belong to Order Orthoptera of class Insecta containing 20,000 described species all over the world. Grasshoppers distinguished by medium or large sized, hind leg modified for jumping, biting and chewing type mouth parts. Due to polyphagus habitat and cosmopolitan distribution have gained world wide importance. The grasshoppers are feed on almost all kinds of plants and skeletonise them, as a result growth of crops and yield adversely affected.

The book has been divided into nine chapters. First chapter is devoted for general introduction that embodies characteristics of Western Ghats, importance of grasshopper diversity. The second topic is devoted to review of literature.

Third chapter deals with materials and methods adapted for work. The grasshoppers have been collected with the help of insect collecting net. Collected fauna have been preserved in insect storage box and identified by consulting appropriate literature. Survey of Grasshoppers has been made by visiting and collecting grasshoppers from the study spots at 15 days interval by one man one hour search method. Distribution records of grasshopper were made by spot observations at different study spots. Grasshopper richness was estimated by sweeping the sweep net for about 30 minute in 15x15 meter area as described by Velez (2008). Collected species identified, counted and morphologically observed. Check list of grasshopper have been prepared.

Forth chapter embodies the taxonomical diversity of grasshoppers. In all, 50 species of grasshopper were redescribed and 3 species newly described. The newly described species refer to *Stauradorus indica* sp. nov., *Pternoschirta sahydricus.* nov. and *Tugsta apludi* sp nov. Fifth chapter is devoted for seasonal abundance, distribution, richness and check list of grasshopper species from Kolhapur, Satara, and Sangli districts of Western Maharashtra. In all, 54 species of grasshopper reported from three districts of Western Maharashtra, out of which 44 species were common and 10 species were rare. The family acrididae contain 5, Tetrigidae 2 and Tettiigoniidae 3 rare species. Rare species have been reported from subfamilies Gomphocerinae, Romalinae, Cyrtacanthacridinae, Eprepocenemidi, Metrodorinae, Tetriginae, Phaneropterinae Oxylakinae and Pseudophyllinae. Sixth chapter devoted to biology of Grasshoppers. The chapter seventh embodies control measures of Grasshoppers. The chapter eighth deals with conclusion. In all, 54 species of grasshopper belonging to four families, 21 subfamilies, and 47 Genera, from Satara, Sangli and Kolhapur district of Western Maharashtra have been studied.

Bibliography

Agabiti, B., Valentionotti, R. and C. Salvadori, 2005. The grasshoppers (Insecta : Orthoptera) of sub-mediterranean zone of the Trentino region (North-East Italy). *Studi. Trent, Sci. Nat. Acta Biol.* 82 : 61-67.

Ahnesjo, J. and A. Forsman, 2006. Differential habitat selection by pygmy grasshopper colormorphs; Interactive effects of temperature and predator avoidance. *Evolutionary Ecology* 20: 235-257.

Alexandre, L., Sward, G., Sergeev, M., Cigliano, M.M. and M. Lecoq, 2011. Locust and Grasshoppers : Behaviour, Ecology and Biogeography. *Hindawi Publishing Corporation*, I.D. 578327.

Ananthaselvi R., Suresh P., Janarthanan S., Karthikeyan K.A. and I. Vijayakumar, 2009. Acridid (Orthoptera) fauna of agricultural ecosystem in some southern district of Tamil Nadu, India. *Journal of Threatened Taxa* 1(9): 491-492.

Andersen, A. N., Ludwia, J.A., Lowe L, M. and C.F. Renz, 2001. Grasshopper biodiversity and bioindicators in Australian tropical Savannas. Responses to disturbance in Kakuda National Park. *Austra Ecology*. 26 : 213-222.

Aswathanarayan, N.V., Ashwath S.K. and K.R. Manjunatha, 1981. On the occurrence of metacentrics in a south Indian grasshopper. *Curr. Sci.* 50(6) : 293-294.

Badenhausser, I., Amouroux, P. and V. Bretagnolle, 2007. Estimating acridid densities in grassland habitats: A comparison between presence–absence and abundance sampling designs. *Environmental Entomology* 36(6): 1495-1503.

Balkrishnan, P. and K.M. Alexander, 1984. The endangered Wildlife of India and their conservation. *Bull. Ethol. Soc. India* (Suppl.) 3, 181-190.

Batary, P., Orchik M., Baldi, A., Kleijn, D., Kisbenedek, T., and Erdos, S. 2007. Effect of local and landscape scale and cattle grazing intensity on orthoptera assembrages of the Hungarian Great Plain. *Basic and Applied Ecology*, 8 : 280-290.

Beason, C.F. 1941. Ecology and control of forest insect of India and neighbouring countries, Govt. of India (1961 Reprint), pp. 767.

Bhoje, P.M. and T.V.Sathe, 2003. Faunistic studies on butterflies from Radhanagari Wildlife Sanctuary, India. *Indian J. Environ. and Ecoplan.*, 7(3), 655-658.

Bhowmik, H.K. 1986. Grasshopper fauna of West Bengal. *Zoological Survey of India*. pp 1:179.

Bhowmik, H.K. 1990. Indian species of the genus *Aulacobothrus* Bolivar (Orthoptera : Acrididae). *Rec. Zool. Surv. India.* 86(3&4): 413-423.

Bhowmik, H.K. and P. Haldar, 1984. Preliminary distribution with remarks on little known species of Acridiidae (Orthoptera : Insecta) from the Western Himalayas (Himachal Pradesh). *Records of the Zoological Survey of India* 18 (1& 2): 167-191.

Bhowmik, H.K. and K.N. Rai, 1984. Notes on a collection of grasshoppers (Orthoptera: Acrididae) from the Siwalik Hills, *Indian Mus. Bull.*, 17:48-54, 2- P15.

Bhowmik, H.K., Saha B.C. and R.N. Bhargava, 1990. Contribution to the Acridid fauna (Orthoptera) of North–Eastern states of India. *Rec. Zool. Surv. India.* 86(2): 217-227.

Bhusnar, A.R. and T.V. Sathe 2011. Diversity of Grasshoppers from Kolhapur district Maharashtra, India. *Geobios* 38:277-282.2.

Bhusnar, A.R. and T.V. Sathe 2012. Ecology of Grasshoppers (Order: Orthoptera) from Kolhapur district, Maharashtra, India. *Bionano Frontier* 5:144-147.

Bieringer, G. and K.P. Zulka, 2003. Shading out species richness: edge effect of Pine plantation on the Orthoptera (Tettigoniidae and Acrididae) assemblage of adjacent dry grassland. *Biodiversity conserve.* 12: 1481-1495.

Bisby, F.A., Y.R. Roskov, M.A. Ruggiero, T.M. Orrell, L.E. Paglinawan, P.W. Brewer, N.J. Bailly, and V. J. Hertum (2007). Species 2000 and ITIS Catalogue of life : 2007 Annual Checklist.

Blackwelder, R.E. 1967. Taxonomy, John Wiley and Sons, Inc., N.Y. 698 pp.

Bolivar, I. 1899. Les Orthopters de St. Joseph's College Trichinopoly (Sudx de I Inde); Ze Partie, Annis Soc. Ent. Fr., 68: 781-810, p. 15. 11-12.

Branson D. H. 2011. Relationships between plant diversity and grasshopper diversity and abundance in the little Missouri National Grassland. *Psyche,* : 1-7.

Branson D. H. and G. A. Sword, 2009.Grasshopper herbivory affect native plant diversity and abundance in a grasshopper dominated by the exotic grass. *Agropyron cristatum. Restoration Ecology.* 17(1) : 89-96.

Branson, D.H. 2011. Relationships between Plant diversity and Grasshopper Diversity and Abundance in the little. Missouri National Grassland. *Hindawi Publishing Corporation* ID-748635.

Brust, M., Hoback, W.W. and R J. Wright, 2007. Immersion tolerance in Rangeland grasshopper (Orthoptera : Acrididae). *Journal of Orthoptera Research* 16(2) : 135-138.

Champion, H.G. and S.K. Seth 1968. A revised survey of the forest type of India. Manager of Publication, Delhi, pp. 464.

Chandra K., Gupta S.K. and M.S. Shishodia 2007. A checklist of Orthoptera of Madhya Pradesh and Chhattisgarh. *Zoo. Print Journal* 22(5) : 2683-2687.

Chandra, K. 2003. Insecta biodiversity in Madhya Pradesh and Chhatisgarh. *Advancement in Insect Biodiversity Agrobios (India) Jodhpur*: 37-52.

Chandra, K. and S.K. Gupta 2009. On a collection of Orthoptera fauna of Veerangana Durgavati Wildlife Sanctuary, Damoh (Madhya Pradesh). *Biological Forum–An International Journal.* 1(1) : 77-82.

Chandra, K., S.K. Gupta and M.S. Shishodia. 2007. A checklist of Orthoptera of Madya Pradesh and Chhattisgarh. *Z00's Print Journal.* 22(5):2683-2687.

Chatterjee, P.N. and M.P. Mishra, 1975. Natural insect enemy and plant host complex of forest insect pests of Indian region. *Indian For. Bull. No. 265* (M.S.) Entomology, Controller of Publications, Govt. of India, Delhi pp. 1-223.

Cigliano, M. M., De Wysiecki, M. L. and C. E. Lange, 2000, Grasshopper (Orthoptera : Acridoidea) species diversity in the Pumpas, Argentina. *Diversity and distribution.* 6 : 81-91.

Cigliano, M. M., Torrusio, S., M. L. De Wysiecki, 2002. Grasshopper (Orthoptera : Acrididae) community composition and temporal variation in the Pumpas, Argentina. *Journal of Orthoptera Research.* 11(2) : 215-221.

Coleman, L.C. and K.K Kannan. 1911. The rice grasshopper (*Hieroglyphus banion* Fabr.). Bull. Dep. Agric. Mysor (Ent.) No. 1:1-52, 5.

Cotes, E.C. 1891: The locust of North–Western India, *Acridium peregranum. Journal Bombay Natural, History Society.* 6 : 242-262.

Darwin, C. 1859. On the origin of species by means of Natural selection. John Marry, London.

Davis, P.H. and V.H. Heywood, 1963. Principles of angiosperm taxonomy Edinburgh; Oliver and Boyd, 558 pp.

De Wysiecki, Maria. L., Sanchez, Norma. E. and Ricci, E. Susana. 2000. Grassland and shrub land grasshopper community composition in northern La Pampa Province, Argentina. *J. Orthoptera Res.* 9 : 211-221.

Dirsh, V.M. 1965. The Arican Genera of Acridoidea. UK : Cambridge University Press.

Dreeze, J. and A. Sen, 1995. Economic development and social opportunity, New Delhi. Oxford University Press, pp. 291-292.

Dwivedi, K.P. 1978. Energy flow of the grasshopper *Conocephalus pallidus* Redt. Tettigoniidae : Orthoptera population in Mohan-Bhata grassland, Madhya Pradesh *Indian Journal of Ecology*, 5(2) : 222-228.

Dwivedi, K.P. 1990. Bioenergetics of the grasshopper *Catantops pinguis innotabilis* (Walker) (Acrididae : Orthoptera) Population in Mohanbhata grassland. *Environmental Ecology* 8 (4) : 1095-1100.

Ehrlich, P.R. and A.H. Ehrlich, (1981). The cause and consequences of the disappearance of species. *Randome House*, New York, 1981.

Ely, S.O., Peter, G.N., Njagi, Bashir, M.D., Salah E1-Tom El-Amin and Ahmed Hassanali, 2011. Diel Behavioural Activity patterns in Adult Solitarious Desert Locust, *Schistocerca gregaria* (Forskal) *Hindawi Publishing Corporation* Article ID 459315.

Floren, A., Riede, K. and Ingrisch, Sigfrid. 2001: Diversity of Orthoptera from Bornean lowland rain forests trees. *Ecotropica.* 7 : 33-42.

Forman, R.T. and M. Godran, 1986. Landscape Ecology John Widey and Jons New York NY.

Forsberg, Bjorn 2004. Microhabitat, Phenology and diversity of Orthoptera in a seminatural pasture. Examensarbetan / seminarieuppsatser, Institutionnen for ekologi och vastproduktionslara Uppsala. 1-22.

Frankel, O.H. 1970. Variation–the essence of life. *Proc. Linn. Soc., New South Wales*, 95 : 158-169.

Gadagkar, R.K. and Chandrasekara and P. Nair. 1990. Insect species diversity in tropics:Sampling methods and a case study. *Journal of Bombay Natural History Society*. 87:337,353.

Gaonkar, H. 1996. The butterflies of Western Ghats India and Srilanka. A biodiversity assessment on the ecological history of Western Ghats Abst., pp. 99.

Gardiner Tim., Hill Julian, 2004: Feeding preferences of *Chorthoppus parallelus* (Orthoptera : Acrdidae). *Journal of Orthoptera Research.* 13(2): 197-203.

Gardiner Tim., Hill Julian and David Chesmore, 2005. Review of the methods frequently used to estimate the abundance of Orthoptera in grassland ecosystems. *Journal of Insect Conservation.* 9 : 151-173.

Gebeyehu, S. and M.J. Samways, 2002. Grasshopper assemblage response to a restored national park (Mountain Zebra National Park, South Africa). *Biodiversity and conservation.* 11 : 283-304.

Gebeyehu, S. and M. J Samways, 2006.Conservation refugium value of a large mesa for grasshoppers in South Africa. *Biodiversity and Conservation.* 15 : 717-734.

Gough, Le. Osenberg, C. W., Gross K.L. and S.L. Collins, 2000. Fertilization effects of species density and primary productivity in herbaceous plant communities. *Oikos.,* 89, 428-439.

Green S.V. 1998. The latonomic impediment in Orthopteran research and Conservation. *Insect conservation.* 2, 151-159.

Gunther, K. 1939. Rivision der Acrydiinae (Orthoptera) III Sectio Amorphopi (Metrodorae Bol. 1887, aul). Abh. Ber. Staatl. Mus. Tierkunde Dresden, Reine A: Zoologie 20 (NF Bd.1):16-335.

Guo, Zhon–Wel., Li. Hong–Chang and Gan, Ya – Ling, 2006. Grasshopper (Orthoptera : Acrididae) biodiversity and grassland ecosystems. *Insect Science.* 13 : 221-227.

Gupta S.K. and K. Chandra, 2009. Orthoptera fauna of Madhav National Park, Shivpuri, Madhya Pradesh. India. *Uttra Pradesh J. Zool.,* 29(1): 79-89.

Hammond, P.M. 1992. Species inventory. In Global Biodiversity, Status of the Earth's Living Resources. Pp. 17-39.

Hawkshworth, D.L. and M.T. Kalin-Arroyo, 1995. Magnitude and distribution of biodiversity. In Global Biodiversity Assessment. Cambridge University Press. pp. 107-191.

Hawksworth, D.L. 1994. Biodiversity measurement and estimation, *Phil. Trans. Roy. Soc., Lon.* B. 345 : 1-136.

Heller, K.G., Orchi, K.M., Grein, G. and S. Ingrisch 2004. The *Isophya* species of Central and Western Europe (Orthoptera : Tettigoniidae : Phaneropteridae). *Tijdschrift voor Entomologie* 147 : 237-258.

Heslop-Harison, J. 1963. Species concepts:Theoretical and practical aspects. In Chemical Plant Taxonomy, *T. Swan. Academic Press inc.,* N.Y., pp.17-40.

Hewit, G.B. 1989. Spatial analysis of the relationship of grasshopper outbreaks to soil classification on : Lecture notes in Statistics 55 : 219-238.

Hewilt G.B. 1985. Review of factors affecting fecundity, oviposition and egg survival of grasshopper in North America-US Dep. Agric. *Agric. Res. Serv.* ARS 36.

Hill, Jovonn. G. 2007: The grasshopper (Orthoptera : Romaleidae : Acrididae) fauna of Black Belt Prairie remnants in Alabama and Mississippi. *Journal of Orthoptera Research.* 16(2) : 139-144.

Hoell, H.V., Doyen J.T. and A.H. Purcell, 1998. Introduction to Insect Biology and Diversity. Oxford University Press pp. 392-394.

Hubbell, T.H. 1932. A revision of the puer group of the North American genus *Melanoplus,* with remarks on the Taxonomic value of the concealed male Genitalia in the Cyrtacanthacrinae (Orthoptera : Acrididae). *The University of Michigan* Press, 23 : 1-64.

Ingrisch, S. 1989. Zur Laubheuschrecken fauna von Thailand (Insecta : Saltoatoria : Tettigoniidae). *Senckenbergiana boil.,* 70(1/3) : 89-138.

Ingrisch, S. (1998): Monograph of the Oriental Agraeciini (Insecta, Ensifera, Tettigoniidae): Taxonomic revision, phylogeny, biogeography, stridulation and development. *Cour. Forsch. Inst. Stencknberg.*206:1-391.

Ingrisch, S. 2002. Orhoptera from Bhutan, Nepal and North India in the Natural History Museum Basel. *Entomologica Basiliensia.* 24 : 123-159.

Ingrisch, S. and A. Gorai, 2001. Orthopteroid insects from Ganesh Himal, Nepal. *Buchreine Zur Entomologie* 8 : 755-770.

Ingrisch, Sigfrid and M.S. Shsihodia, 1997. A new species of Agraeciini from Northeast India (Ensifera : Tettigoniidae). *Entomol. Z.* 107(12) : 510-512.

Ingrisch S. and M.S. Shishodia, 1998. New species and records of Tettigoniidae from India (Ensifera). *Bulletin De La Societe Entomologiae Suisse* 71, 355-371.

Ingrisch, S. and M.C. Muralirangan, 2003. A new species of Himertula (Orthoptera : Tettigoniidae) and Additional Records of Tettigoniidae from Tamil Nadu (India) Bonner. *Zoologische Beitrage.* 51 (4): 305-312.

Isely, F.B. 1944. Correlation between mandibular morphology and food specificity in grasshopper. *Annals of Entomological society of America.* 37:47-67.

Jadhav, B.V. and T.V. Sathe, 2006a. Biodiversity of aphids (Order–Hemiptera) from Satara district of Western Ghats. *Indian J. Environ. and Ecoplan.*, 12(1), 237-240.

Jadhav, B.V. and T.V. Sathe, 2006b. Biodiversity of aphids (Order–Hemiptera) from Poona district of Western Ghats. *J. Adv. Zool.*, 27(1), 43-45.

Jago, N. D. 1998. The world–wide magnitude of Orthoptera as Pests. *J. Orthoptera Research.* 7: 117-124.

Jago, N.D. and H.K. Bhowmik, 1990. A new species of Oedipodine grasshopper, *Sphingonotus orissaensis,* from Eastern India (Orthoptera : Acrididae). *Rec. Zool. Surv. India.* 87 (3): 193-196.

Jenkins, M. 1992. Global Biodiversity status of the Earth's Living Resources. Chapman and Hall, London, pp.40-46.

Joern, A. 1975. Grasshopper affinities and habitat relations in the Solitario, 121-123.

Joern, A. 1982. Distribution, Densities and relative abundances of grasshoppers (Orthoptera : Acrididae) in a Nebraska Sandhills Prairie. *The Prairie Naturalist.* 14(2) : 37-45.

Joern, Anthony 2005. Distribution by fire frequency and Bison grazing modulate grasshopper assemblages in Tallgrass Prairie. *Ecology.* 86(4) : 861-873.

Johnson D.L. 1989. Spatial analysis of the relationship of grasshopper outbreaks to soil classification lecture notes in statistics, 55 : 219-238.

Jonas, J. L. and A. Joern, 2007. Grasshopper (Orthoptera : Acrididae) communities respond to fire, bison, grazing, and weather in North American tallgrass prairie : a long-term study. *Oecologia.* 153 : 699-711.

Joseph, K. J. 1984. Insect life and ecodevelopment of the Western Ghats.

Joseph, T.M. 2004. Biodiversity conservation in the Western Ghats. Biodiversity and Environment, 17, 221-227.

Julka, J.M., S.K. Tandon, P. Halder and M.S. Shishodia 1982. Ecological observation on the grasshoppers (Orthoptera–Acridoidea) din Solan (H.P.) India. *Oriental Insects* 61 : 63-71.

Kanade, M., Raguraman, S., Ganapathy, N., and K. Gunathilagaraj, 2004: Orthopteran diversity in irrigated rice ecosystem in Madurai, Tamil Nadu. *Zoo's Print Journal.* 19(10): 1663-1664.

Kanade, R., J., Taswalkar. J., Shukla, S., Champhekar, K., Bhawalkar, S., Mone, S., Raghavendra, S., Chandrasekar, M., Sardesai, M., Kushalappa, C. and A. Patwardhan, (2008): Vegetation survey of Chandoli National Park, from Northern Western Ghats of India. *J. Econ. Taxon. Bot.* 32 (4) : 930-938.

Kandibane, M., Raguraman, S., Ganapathy N. and K. Gunathilagaraj, 2004. Orthoptera diversity in irrigated rice ecosystem in Madurai, Tamil Nadu. *Zoo's Print Journal.* 19(10) : 1663-1664.

Kavane, R.P. and T.V. Sathe, (2011). Wild Silk technology. *Daya publishing house, New Delhi*, pp. 1-234, ISBN- 948-81-7035-712-4.

Kemp, W.P., Harvey, S.J. and K.M. O'Neill, 1990. Patterns of vegetation and grasshopper community composition. *Oecologia.* 83 : 299-308.

Keystone Center, 1991. Final consensus Report of the Keystone Policy Dialogue on biological Diversity on Federal Lands. The Keystone center, keystone, Brandon, Canada.

Khalid Mahmood, Maqsood Ahmad and Sajjad Anwar, 2004. Tetrigidae (Orthoptera) of Azad Jammu and Kashmir. *Pak Emntolo,* vol. 26, 1

Kim Tae- Woo and Kim Jin, 2004. A taxonomic study of Korean Tetrigidae (Orthoptera : Caelifera : Tetrigidea). *Entomological Research.* 34(4) : 261-267.

Kirby, W.F. 1914. The fauna of British India Ceylon and Burma. Orthoptera (Acridiidae). London IX 276 pp.

Krausz, K., Papai J. and L. Galle. 1995. Composition of Orthoptera assemblages in grassland habitats at Lower–Tisza flood plain–Tiscia 29, 47-52.

Krishnamurthy, K.V. 2004. An Advanced Textbook on Biodiversity Principals and Practice. *Oxford and IBH Publishing.*, New Delhi, pp. 1-266.

Kruess, Andreas and Teja Tscharntke, 2002. Grazing intensity and the diversity of Grasshopper, Butterflies and trap-nesting Bees and Wasps. *Conservation Biology* (16)6 : 1570-1580.

Kudla, M.L., D.E.Wilson and E.O.Wilson (1971). Biodiversity II Understanding and protecting our Biological Resources. Joseph Hennry, Washington, DC.

Lanjar, A. G., Talpur, M. A., Khuhro, R. D. and H. Q. Khalid, 2002. Occurrence and Abundance of Grasshopper species on Rice. *Pakistan Journal of Applied Sciences.* 2(7) : 763-767.

Larson, T.B. 1987. The butterflies of the Nilgiri mountains of South India (Lepidoptera : Rhopalopcera). *J. Bombay Nat. Soc.*, 84(1), 26-54; 84 (12), 291-316; 84 (3), 560-584.

Latchininsky, Alexandre., Sword, Gregory., Sergeev, Michael., Cigliano, Maria. Marta and Michel Lacoq, 2011. Locust and grasshopper : Behaviour, Ecology and Biogeography. *Psyche.* 578327 : 1-4.

Lecoq, Michel 2002. Deseart locust management : from ecology to anthropology, *Journal of Orthoptera Research* 14(2) : 179-186.

Lecoq, M., Chamouine A. and My-Hanh, Luong–Skovmand 2011. Phase–Dependent Phase Polyphenism in Nonmodel Locusts: A Minireliew, Hojun Song. *Hindawi Publishing Corporation* Article ID 105352

Lecoq, Michel., Chamouine, Abdou and Luong–Skovmand, My – Hanh, 2011. Phase dependent color polyhenism in field population of Red Locust Nymphs (*Nomadacris septemfasciata* Serv.) in Madagascar. *Psyche.* 1-12.

Liu, Chun-Xiang and Kang Le 2009. A new genus, *Paraxantia* gen. nov., with descriptions of four new species (Orthoptera : Tettigoniidae : Phaneropterinae) from China. *Zootaxa.* 2031 : 36-52.

Lockwood, J.A. 1990. Rangeland grasshopper ecology. *The bionomics of grasshoppers, katytids and their kin*. UK pp. 83-1202.

Lockwood, J.A. 2010. The Onthology of Biological Groups DO grasshoppers from Assemblages, Communities guilds, populations or something else *Hindawi Publishing Corporation. Psyche.* 501983: 1-9.

Lockwood, D.R. and J.A. Lockwood, 2008. Grasshopper population ecology : Catastrophe, Criticality and Critique. *Ecology and Society* 13 (1) : 34.

Lockwood, J. A., and M. G. Sergeev, 2000. Comparative biogeography of grasshoppers (Orthoptera : Acididae) in North America and Siberia : Application to the conservation of biodiversity. *Journal of Insect Conservation.* : 161-172.

Mandal, S.K., Dey, A. and A.K. Hazra, 2007. Pictoral Handbook on Indian Shorthorned grasshopper pests (Acridoidea : Orthoptera). *Zool. Sarv, India Kolkata*, 1-57.

Mani, M.S. 1994. General Entomology. Oxford and IBH publishing Co. Pvt. Ltd., Calcutta. ISBN-81-204-0386-X.

Marini, Response., Fontana, Paolo., Bhattisti, Andrea and K. J. Gaston, 2009. Response of Orthopteran diversity to abandonment of semi-natural meadows. *Agriculture, Ecosystems and Environment.* 1-5.

Mason, H.L. 1950. Taxonomy, systematic botany and biosystematics. Madrono, 10 : 193-208.

Mass, B. 2009. New and less known Orthoptera (Insecta from the island of Socotra (Yemen.). *Zoo. Taxa* 2132 : 53-64.

Massa, Bruno, 2009. Annotated check-list of Orthoptera of Libya. *Journal of Orthoptera Research.* 18(1) : 75-93.

Massa, Bruno, 2009. New and less known Orthoptera (Insecta) from the island of Socotra (Yemen). *Zootaxa.* 2132: 53-64.

Mathew Goerge., Shamdudeen, R.S. and C. M. Brijesh, 2007. Insect fauna of Neyyar Wildlife Sanctuary, Kerala, India. *Zoo's Print Journal.* 22(12) : 2930-2933.

May, R.M. 1995. Conceptual aspects of the quantification of the extent of biological diversity. *Phill. Trans. R. Soc. London Ser. B.* 345, 13-20.

Mayya, S. Sreepada, K.S. and Hegade, M., Jayarama, 2003. Survey of short horned grasshoppers (Acrididae) from Dakshina Kannada district, Karnataka. *Zoos. Print Journal.* 20(9) : 1977-1979.

Mehta, H.S., V. K.Mattu and S.K. Thakur (2002). Orthopteran diversity of Kalatop-Khajjiar Wild Life Sanctuary, Chamba, Himachal Pradesh. *Bionotes*, 4(3):60.

Muhammad Azhar., Suhail Anjum., Sabir Arshed Makhdoom and Saeed Asif, 2000. Biosystematic studies of Genus *Formosotettix tinkham* (Tetrogodae : Orthoptera) of Paddy Tract of the Punjab, Pakistan. *Pakistan Journal of Biological Science.* 3(9) : 1530-1531.

Muralirangan, C. 2002. Effect of abiotic factors on the population of an acridid grasshoppers, Diabolocatantops pinguis (Orthoptera : Acriididae) at two sites in Southern India : a three year study. *Journal of Orthoptera Research*, 11(1):55-62.

Mukherji, M.K. and R.L. Randell 1975. Estimation of embryonic development in populations of *Melanoplus sanguinipes* (Fab.) (Orthopera : Acrididae) in the fall. Acrida 4 : 9-17.

Mukhtar, Gule–e- Shahdab., Nawaz, Mohammad., Nawaz, Yasmin and Kakar, Asmathullah, 2010. Biodiversity and occurrence of Grasshopper (Acrididae : Orthoptera) of Quetta division Balochistan. *Pakistran. J. Zool.* 42(1): 87-91.

Mulkern,G.B. (1967). Food selection by grasshoppers. *Ann Rev. Entomol.*, 12, 59-78.

Nagendran, N.A. and M.S. Smija 2004. A seminal report on diversity of higher taxa of aquatic insects in three hills streams of Western Ghats, South India.

Nagy Antal., Solymos Peter and I. A., Racz, 2007. A test on the effectiveness and selectivity of three sampling methods frequently used in Orthopterological field studies. *Entomologia Fennica.* 18 : 149-159.

Nair, K.S. and Mathew, 1993. Diversity of insects in Indian forests. The state of our knowledge. *Hexapoda*, 5 (2), 71-78.

Nair, K.S., Mathew, G. and M. Shivarajan 1973. Occurrence of bagworm Pterowea plagiophleps Hampsen (Lepidoptera : Psychidae) as a pest of the free Albizia falcataria in Kerala, India. *Entomon.*, 6, 179-180.

Noss, R.F. 1992. Issues of scale in conservation biology In : fiedler, P.L. and Jain S.K. (Eds.) Conservation Biology. Champman and Hall, New York, NY, pp. 239-250.

Noss, R.F. 1996. Conservation of Biodiversity at the landscape scale. In : Szaro, R.C., and Johnston, D.W. (Eds.) Biodiversity in Managed Landscapes Oxford Univ. Press, Oxford, pp. 574-589.

Nufio, Cesar. R., Mc Clenahan, Jeff. L. and Bowers, M. Deane (2011) : Grasshopper response to reductions in habitat area as mediated by subfamily classification and life history traits. *J. Insect Conservation.* 15: 409-419.

Ock, J.E. 2004. The structure and function of Auditory chordotonal organs in insects. *Microscopy Research and Technique* 63 : 315-337.

Olfert, O., Weiss, R.M. and Kriticos, D. 2011. Application of General circulation Models to Assess the potential Impact of climate change on potential Distribution and Relative Abundance of *Melanoplus sanguinipus* (Fabricius) (Orthoptera : Acrididae) in North America. *Hindawi Publishing Corporation* Article ID 980372.

O'Neill, Kevin, M., Larson, Deanna, P. and W. P, Kemp, 2002. Sweep sampling technique affect estimate of the relative abundance and community composition of grasshopper (Orthoptera : Acrididae). *J. Agric. Urban Entomol.* 19(3): 125-131.

Paranjpe, S. Y. and A. M. Bhalerao, 1984. Biological observations on a pigmy locust; *Potusa sabulosa* Hankock. Tetrigidae : Orthoptera)., 92: 331-336.

Pandharbale, A.R. 2004. Biodiversity of Moths (Lepidoptera) from Western Ghats of Satara district (Maharashtra). Ph. D. Thesis, Shivaji University, Kolhapur. pp 1. 240

Pandharbale, A.R. and T.V. Sathe, 2001. On a new species of the Genus *Syntomis* (Syntomidae : Lepidoptera) from the environment of Western Ghats (Satara District). *Indian J. Environ. and Ecoplan.,5(2), 601-602.*

Paranjape, S.Y. and A.M. Bhalerao, 1984. Biosystematic and Ethoecological studies on family Tetrigidae (Orthoptera). *Oriental Entomology Symposium, Trivandrum Proceedings.* 19-23.

Paranjape, S.Y. and A.M. Bhalerao, 1984. Bioecological observations on a pigmy locust, *Potua sabulosa* Hancock (Tetrigidae : Orthoptera). *Psyche,* 92 : 331-336.

Paulraj, M.G., V. Anabalagan and S. Ignacimutha, 2009. Distribution of grasshoppers (Insecta : Orthoptera) among different host plants and habitats in two districts of Tamil Nadu; India. *Journal of Threatened Taxa* 1 (4) : 230-233.

Phipps, J. 1970. Notes on the biology of grasshoppers (Orthoptera : Acridoidea) in Sierra Leane, *J. Zool. Lond.*, 161 : 317-319.

Pisica E. I., Iorgu I. S. and M. Dumitra, 2008, Preliminary data regarding the Orthoptera (Insecta : Orthoptera) specific diversity from Bucharest Metopolitan area. *Analele stiintifice ale Universitatiices. Al. I Cuza, Iasi, B. Biologie animal, Tom LIV.*

Preveling, Ralf, 2001. Environmental conservation and locust control– possible conflicts and solutions. *Journal of Orthoptera Research.* 10(2) : 171-187.

Priya, A. V. and T.C. Narendran, 2003. A key and a checklist of the genera of short-horned grasshopper (Orthoptera : Acridoidea) of Kerala. *Entomon.* 28(3): 223-230.

Pungis, V. 2007. Fauna and Ecology of grasshoppers (Orthoptera) in the coastal Dune habitats in Ziemupe nature Reserve, Cativo. *Latvijas entomlogs*, 44:58-68.

Radhakrishnan, L. 2003. India's environmental Problems and Issues. In Environment and its challenges (Ed. Arvind Kumar) 13, 34-38.

Rentz, Def., Su, You. Ning and Ueshima, Norihiro 2009. Studies in Australian tettignoniidae : The phyllophorinae (Orthoptera : Tettigoniidae : Phyllophorinae). *Zootaxa.* 2075 : 55-68.

Rodger, W.A. and H.S. Panwar, 1988. Planning of Wildlife Protected Area Network in India-1 and 2, DehraDun : Wildlife Institute of India.

Roonwal, M.L. (1981). Field bioecology and morphometry of some central Indian grasshoppers (Acridoidea) with notes on a swimming species (Tetrigoidea). Prolledings of Zoological Society of Calcutta 32 : 97-106.

Roonwal, M.L., G.D. Bhasin and G.D. Pant, 1950. A systematic catalogue of the main identified entomological collection at the Forest Research Institute, Dehra Dun. Parts 1-3. Indian For., 76(11), 498-505.

Rowell, C. H. 1998. The grasshopper of Costa Rica : a survey of the parameters influencing their conservation and survival. *Journal of Insect Conservation.* 2 : 225-234.

Samways, M.J. 1994. Insect conservation. Biology, London : Champman and Hall.

Samways, M. J. and K. Kreuzinger, 2001. Vegetation, ungulate and grasshopper interactions inside vs. outside and African savanna game park. *Biodiversity and Conservation.* 10 : 1963-1981.

Sathe, T.V. 1992. Fauna of aphids on plants of economic importance found in Western Maharashtra, India. *J. Curr. Biosci.* 9(1), 27-31.

Sathe, T.V. 2004. Biology and behaviour of coccinellid beetles, In Indian insect predators in biological control Ed. K. Shayaraj, 8, 177-198.

Sathe, T.V. and M.K. Mulla, 1995. Insect pests of Mulberry from Amboli. *Oikoassay,* 12 (1-2), 9.

Sathe, T.V. and A.R. Pandharbale, 1999. Hawk moth (Sphingidae : Lepidoptera) from Western Ghats of Satara district, India. *Bull. Biol. Sci.* : 1, 81-88.

Sathe T.V. and A.R. Pandharbale, 2005. Biodiversity of Moths (Order–Lepidoptera) from Western Ghats of Satara district, India. *Bull. Biol. Sci.* 1, 81-88

Sathe, T.V. and K.P. Shinde, 2006. Diversity of butterflies from Western Ghats (Kolhapur district). *J. Nat. Con.,* 81(1), 181-184.

Sathe, T.V. and Shinde K.P. 2008. Dragonflies and pest management. *DPH.* New Delhi. pp.1-179.

Sathe, T.V. Inamdar S.A. and M.V. Santhakumar, 1986-87. Fauna of butterflies from Western Maharashtra and Western Ghats (Part of Maharashtra only). *Indian J. Shivaji Uni. (Science),* 23, 391-398.

Satish, P.M. 1996. Moths and butterflies of Bhadra Projects. M.Sc. Thesis submitted to Kumempu University, Shimoga, pp. 80.

Sazaro, R. and Shapiro, B. 1990. Conserving our Heritage : America's Biodiversity. *The Nature conservancy,* Arlington, VA.

Schmitz, O. J. 2005. Scaling from plot experiments to landscapes: studying grasshoppers to inform forest ecosystem management. *Oecologia,* 145: 225-234.

Senthilkumar N. 2010. Orthopteroids in Kaziranga National Park, Assam, India. *Journal of Threatened Taxa* 12 (10). 1227-1231.

Senthilkumar, N; Nizara B. and N.J. Borah, 2006. Orthopteran fauna of the Gibbon Wildlife Sanctuary, Assam.*Zoo, Pri. J*, 21 (8): 2347-2349.

Sergeev, M.G. 1998. Conservation of orthopteran biological diversity relative to landscape change in temperate Eurasia. *J. Insect Conserv.* 2 : 247-252.

Sergeev, M.G. 2011. Distribution patterns of Grasshoppers and their Kin in the Boreal Zone. *Psyche.* 324130 : 1-9.

Serville, J.G. 1839. In Reret, collection des suites a Buff on Orthopteres. Histoire naturelle des Insects :1-776, pls :1-14.

Sharma R.M., P.P. Kulkarni and R.H. Kamble 1991. Population out burst of *Mecopoda elongata* Linn. (Orthoptera : Tettigoniidae) Around Pune, Maharashtra.

Shinde, K.P. 2008. Biodiversity of dragonflies (Order–Odonata) from Koyana dam and around area. Ph.D. Thesis, Shivaji University, Kolhapur pp. 1-275.

Shishodaia, M.S. 2000. Otrhoptera (Insecta) fauna of Andaman and Nicobar Islands. *Rec. Zool. Surv. India.* 98(3) : 1-24.

Shishodia, M.S. Mehta H.S., Mattu V.K. and S.K. Thakur, 2002. Orthoptera (Insecta) from Pong Dam Wetland, District Kangra (H.P.) India. *Zoos Print Journal,* 18(3) : 1047-1048.

Shishodia, M.S. 1991. Grouse locust (Insect : Orthoptera : Tetrigidae) of Kanha National Park, Mahdola, Madhya Pradesh, India. Records of the Zoological Survey of India 89 (1-4) : 101-104.

Shishodia, M.S. 1991. Tetrigidae of North Eastern India, *Zoological Survey of India* : 1-203.

Shishodia, M.S. 1995. Insecta : Orthoptera fauna of conservation areas 6 : Indravati Tiger Reserve. Zoological Survey of India 11-17.

Shishodia, M.S. 1999. Orhoptera fauna of Patalkot, Chhindwara; Madhya Pradesh, India. *Records of the Zoological Survey of India*, 97(4) : 34-43.

Shishodia, M.S. 2000. Short and long horned grasshoppers and crickets of Bastar district, Madhya Pradesh, India. *Records of the Zoological Survey of India,* 98(1) : 27-80.

Shishodia, M.S. 2006. On a collection of grasshoppers by Dr. M.L. Roonwal from the erstwhile Rewa Estate, Madhya Pradesh. *Bionotes* 8(1) : 11-12.

Shishodia, M.S. and S.K. Gupta, 2009. Check-list of Orthoptera of Himachal Pradesh. *Journal of Threatened Taxa.* 1(11) : 569-572.

Shishodia, M.S. and S.K. Mandal, 1990. New records of Orthoptera (Insecta) from the Nagarjuna Sagar Tiger Reserve Forest, Andhra Pradesh, India. *Rec. Zool. Surv. India.* 87(1) : 65-76.

Shishodia, M.S., H.S. Mehta, V.K. Mattu and Thakudr. (2002). Orthoptera (Insecta) from Pong dam Wetland, District Kangra, Himachal Pradesh, India. *Zoos. Print Journal* 18(3):1047-1048.

Shishodia, M.S. and Sunilkumar Gupta, 2009. Checklist of Orthoptera (Insecta) of Himachal Pradesh, India. *Journal of Threatened Taxa* 1(11): 569-572.

Shivaramkrishnan, K.G., Venkataraman, K., Morthy R.K. Utkash, G. and K.A. Subramanian, 2000. Aquatic insect diversity and ubiquity in the streams of the Western Ghats, India. *J. Indian Inst. Sci.*, 80, 537-552.

Shivaramkrishnan, K. G.; Sridhar, S. and K. Venkataraman, 1990. Habitat, microdistruction, life cycle patterns and tropic relationship of Mayflies of Cardman hills, Western Ghats. *Hexapoda*, 11, 59-64.

Showler, Allan. T. 2003. The importance of armed conflict to Desert locust control. 1986-2002. *Journal of Orthoptera Research.* 12(2) : 127-133.

Singh, A. and A.N. Sadhu, 1986. Agricultural problems in India. Jammu University, Jammu, pp. 386.

Singh, K.J. and O.P. Singh 1992. *Cyrtacanthacris tatarica* Linn. as a pest of soybean pods, *Indian Journal of Plant Protection,* 20(1) : 116-117.

Smith, A. 1776. Wealth of Nation- An Inquiry into the Nature and causes of Wealth of Nation. Strahan and Cadell, UK.

Smith, T. R and Capinera, J. L. 2005. Mandibular morphology of some Floridian grasshopper (Orthoptera : Acrididae). *Florida Entomologists.* 88(2) : 204-207.

Smith, T. T., Froeba, J. G. and Capinera, J. L. 2004. Key to the grasshoppers (Orthoptera : Acrididae) of Florida. *Florida Entomologist.* 87(4) : 537-550.

Song, Hojun. 2006. Description of *Schstocerca cohni* n. sp. and redescription of *s. Socorro* (Dirsch) (Orthoptera : Acrididae : Crytacanthacridinae) from Mexico. *Zootaxa.* 1150 : 43-52.

Song, H. 2011. Density–Dependent phase polyphenism in Non model Locusts: A minireview. *Hindawi Publishing Corporation* Article ID 741769.

Spungis, V. 2007. Fauna and Ecology of grasshoppers (Orthoptera) in the coastal Dune Habitats in Ziemuee Nature Reserve, Latvia. *Latvijas entomologs,* 44 : 58-68.

Spungis, Voldemars 2007. Fauna and ecology of grasshopper (Orthoptera) in the Costal Dune habitats in Ziemupe nature reserve. *Lativia. Latvijas entomologs.* 44: 58-68.

Stabbing, E.P. 1940. Indian forest insects of economic importance–Coleoptera. J.K.J. Brothers, Bhopal pp. 1-648.

Steck, C. E., Burgi, M., Bolliger, J., Kienast, F., Lehmann, A. &Y. Gonseth, 2007. Conservation of grasshopper diversity in a changing environment. *Biological conservation.* 138 : 360-370.

Stewart, David, 1998. Non-target grasshoppers as indicators of the side-effects of chemical locust control in the Karoo, South Africa. *Journal of Insect Conservation.* 2: 263-276.

Stork, N.E., Samways, M.J. and H.A. Eeley, 1996. Inventorying and monitoring biodiversity. *Trends Ecol. Evol.* 11, 39-40.

Streett, D.A. and M.R. McGuire 1990. Pathogenic diseases of grasshoppers. *Biology of Grasshoppers,* New York USA pp. 483-516.

Thakur,. S.K., M.S. Shishodia, H.S. Mehta and V.K. Mattu (2004). Orthopteran diversity of Roper Wetland Panjab, India. *Zoos Print Journal,* 19(11):1697.

Suhail, A., Jalal Ariy, M. and Suhail, G. 1999. Taxonomic studies on Hemiacridinae (Acrididae : Orthoptera) of Pakistan. *Int. J. Agri. Biol.* 1 : 142-144.

Suhail, A., Suhail, G. and Khalid AL-Hariri M. 2001. Some Acridid grasshoppers belonging to sub family cyrtacanthacridinae (Acrididae: Orthoptera) from Pakistan. *Int. J. Agri. Biol.* 3 : 43-46.

Szaro, R.C. and H. Salwasser, 1991. The management context for conserving biological diversity. 10th World Forestry Congress, (Paris, France, Sep. 1991). Revae Forestiere Franaise Actes 2: 530-535.

Tandan, S.K., M.S. Shishodia and A. Dey, 1995. Insecta : Orthoptera, pp. 19-25 In : 7 Fauna of Kanha Tiger Reserve. *Zoological Survey of India.*

Tandon, S.K., Hazra, A.K. and S.K. Mandal, 1988. Observations on the field biology and ecology of some grasshoppers (Orthoptera : Acridoidae) Near Calcutta. *Rec. Zool. Surv. India.* 85(2) : 301-318.

Tandon, S.K., M.S. Shishodia and D. Dey, 1976. On a collection of Orthoptera (Insecta) from Kanha National Park, Mandla, Madhya Pradesh, India, *Newsletts. of Zoological Survey of India* 2(4) : 167-170.

Taylor, R.W. 1975. Submission to the inquiry to the impact on the Australian Environment of the Current Woodchip industry Program. Senate Hansard : transcripts of report to the Australian Standing Committee on Science and Environment. Canberra.

Thakur, S.K. and Thakur, M.S. 2011. Long-horned grasshoppers (Orthoptera : Ensifera) diversity from Sukhana Wildlife Sanctuary Chandigarh, India. *Int. Jr. of Agril. Env. And Biotech.* 4(2) : 119-120.

Thakur, S.K., Shishodia, M.S., Mehta, H.S. and V.K. Mattu, 2004. Orthopteran Diversity of Roper Wetland Punjab, India. *Zoos Print Journal,* 19(11) : 1697.

Trevick, Steve and Morris, Simon 2008. Diversity and taxonomic status of some New Zealand grasshoppers. *DOC Research and Development Series.* 290 : 112.

Uvaraov, B.P. 1925. Grasshoppers from the Mount Everest, Subfamily Acridinae, *Ans. Mag. Nat. Hist.* (9) 16 : 165-173.

Uvarao, B.P. 1955. The aridity factor in the ecology of locust and grasshopper of the Old World. United Nations Educational. Scientific and Cultural Organisation. : 1-37.

Velez, V.M. 2008. Grasshopper biodiversity among three sites with different herbivory regimes in the National Bison Range. UNDERC WEST 2008.

Walker, F. 1871. Catalogue of the specimens of Dermaptera and Saltatoria in the collection of the British Museun part Tettigidae and supplements to earlier parts :(4)811-850.

Waloff, N. and Popov, G.B. 1990. Sir Boris Warov (1889-1970) : The father of acidology. *Ann. Rev. Entomol.* 35, 1-24.

Warov, B.P. 1925. Insects, Orthopter, Acrididae. In mission any Babault dans less provinces centrals de I Inde et drans la Region Occidental de I Himalaya, 1914, Paris : 40 pp.

Willemse, F. and S. Ingrich 2004. A new genus and species of Acrididae from South India (Orthoptera : Acridoidea) *Tijdschrift voor Entomologia* 147 : 191-196.

Willemse, Fer and Ingrisch, Sigfrid 2004. A new genus and species of Acrdididae from South India (Orthoptera : Acridoidea). *Tijdschrift Voor Entomologie.* 147 : 191-257.

Wilson, E.O. 1988a. The diversity of life In : De Blij H.J. (Ed.) Earth's 88 : Changing Geographic Perspective National Geographic Society, Washington DC pp. 68-78.

Wilson, E.O. 1988b. The current state of biological diversity In : Wilson, E.O. and Peters, F.N. (Eds.). Biodiversity National Academy Press, Washington DC, pp. 3-18.

Yadav, S.R. and M.B. Bachulkar 1995. New Plant Record from Satara district–Part I.

Yadav, S.R. and M.M. Sardesai 2002. Flora of Kolhapur District. Pub. *Shivaji University Kolhapur, India* pp. 1-680.

Yack, J.E. 2004. The structure and function of Auditory Chordotonal Organs in insects. *Microscopy Res. and Tech.* 63: 315-337.

Yin, Hong and Yin Xiang-Chu, 2005. Description of two new species of *Stemocatantops* (Orthoptera : Acrididae : Catantopinae) from Taiwan with key to known species of genus. *Zootaxa.* 1055 : 41-48.

Zahn, A., Juen, A., Traugott, M. and A. Lang. 2007, Low density cattle grazing enhances arthropod diversity of Abanodoned Wetland. *Applied Ecology and Environment Research.* 5(1): 73-86.

Zoological Survey of India. 1983. Threatened Animals of India, Calcutta:Zoological Survey of India pp. 307.

Index

www.ingramcontent.com/pod-product-compliance
Lightning Source LLC
Chambersburg PA
CBHW021436180326
41458CB00001B/295

* 9 7 8 9 3 5 1 3 0 1 9 2 9 *